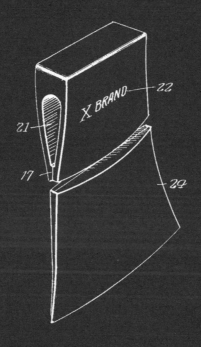

X BRAND

22

21

17

24

Inventors
Howard Adams Vaughan
Gunar Olson and
By Erick Erickson
Fred Gerlach their Atty.

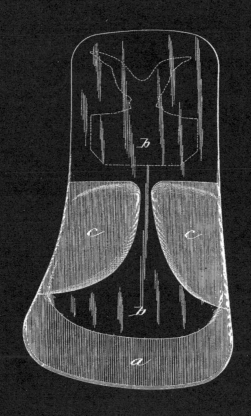

INVENTOR

James P. Kelly,

BY

Chester Bradford,

ATTORNEY.

Buchanan-Smith's
Axe Handbook

HOFFMAN BLACKSMITHING
NEWLAND, NORTH CAROLINA

US FOREST SERVICE
SACO RANGER STATION
CONWAY, NEW HAMPSHIRE

DEAD
END

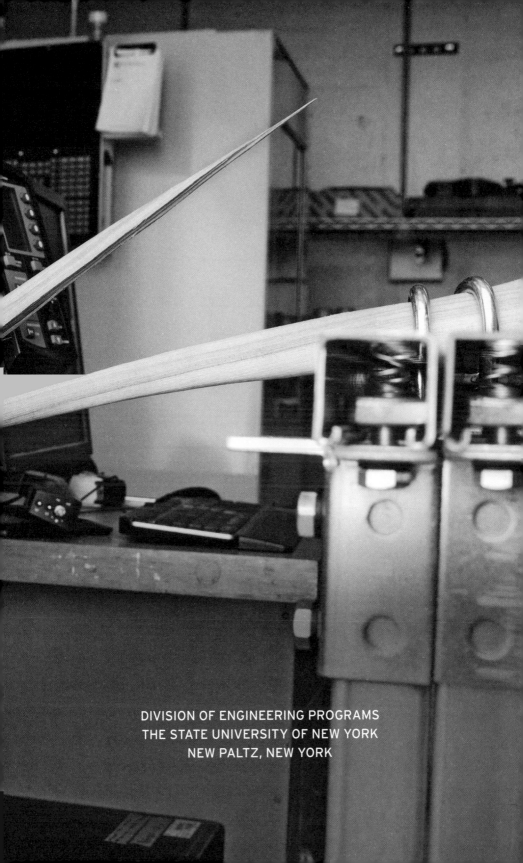

DIVISION OF ENGINEERING PROGRAMS
THE STATE UNIVERSITY OF NEW YORK
NEW PALTZ, NEW YORK

THE AXE MUSEUM AT GRÄNSFORS BRUKS
NORTHERN HÄLSINGLAND, SWEDEN

CATSKILL MOUNTAINS
ANDES, NEW YORK

Buchanan-Smith's

AXE
Handbook

Peter Buchanan-Smith

With Ross McCammon, Nick Zdon,
and Michael Getz

Photography by the author

Abrams Image, New York

CONTENTS

Very humble objects, like old tools, had the direct elegance
generated by a culture sensitive to the requirements of that
tool and its user. Style is the way things, ideas, attitudes take
form. Style is the tangible aspect of intangible things.

—Massimo Vignelli

To my mother (the artist)
and my father (the scientist)

Foreword

This book is a love letter to my favorite tool. It's the
culmination of my personal learnings, so it's purposefully
subjective. That's why I called it *Buchanan-Smith's Axe
Handbook* and not *The Axe Handbook*. It's a reflection of
the lore, history, and best practices of this surprisingly
mysterious tool as I have come to understand it. Axes are
deeply personal: to me, to the contributors of this book, to
Robert Frost (page 102), and to a growing number of so
many enthusiasts. I hope I can help make it personal for
you, too.

Introduction

I recall my father's felling axe, a fixture of the family farm in Canada, where I was born and raised. I remember its shapely curved helve with a fawn's foot handle dipped in bright yellow paint for easy spotting if it was ever left in the long grass or deep snow. I remember the axes from the canoe trips I took at Ahmek, a historic boys' camp in Algonquin Park in northern Ontario. We'd paddle and portage for weeks, equipped with not much more than maps, paddles, cedar-strip canoes, outsized green duck-canvas "turtle" packs with leather straps. We loaded them with sleeping bags packed in black garbage bags and then lashed an axe to the side. I recall the guide, Dave Conacher, a legend at my camp who taught us that with an axe you could do just about anything in the wilderness, and that without it you were done for.

My first axe

I recall the nameless axe that I bought on eBay in 2009. It was an elegant old felling axe with a well-worn helve and a head with a patina that had stories to tell. Years later, I brought it up to the cabin I'd eventually purchase in the Catskill Mountains. But for months it rested in a corner of my workshop in New York City. An object of beauty and

An object of beauty and utility

utility. An emblem of simplicity. From its corner, that axe saw me through a divorce, the death of my dog, and my biggest career change. It saw me say goodbye to all that, and was there when I planted the first seeds of a special sideline project that would grow to become my life's work. In fact, it inspired it.

Best Made Company

Best Made Company started in my garage. It was just me, a few cans of bright enamel paint, a tub of marine spar varnish, and a dozen axes. Best Made came to be known for its bright, bold, graphic, colorful axes—axes made to my specifications by Council Tool, a third-generation forge on the shores of Lake Waccamaw, North Carolina. We didn't just promote and sell axes. We told stories about them. We showed our customers how to use them and restore them. In our catalogs, we showed them being swung in the wild places where they belonged (even if some of our customers were only hanging them in their Tribeca apartments). We shipped thousands of axes to every conceivable corner of the world. Before Best Made, axes were most often relegated to a toolshed. Now they were hanging in art galleries, museums, and boardrooms of Fortune 500 companies. The Best Made product line grew into the hundreds—from axes to first-aid kits to toolboxes to chambray shirts to high-performance down jackets—and in less than ten years we were a full-fledged retail brand with thriving online and catalog sales and stores in New York and Los Angeles.

The axe is embedded in our DNA

Working the retail floor at Best Made, I'd watch customers pick up an axe for the first time. There was always that quiet awe. They'd remark on its weight and sharpness. Sometimes they were speechless. I think that's because the axe—unlike any other tool, maybe any other object—is deeply embedded in our DNA. It's the oldest tool known to humankind, and when you hold one, let alone swing one, you're connecting to an ancient muscle memory.

I left Best Made in part to write this book; it has been many years in the making. To tell the story of the axe, I had to listen to the stories of others. So I rounded up axes from all over the world; I scoured the flea markets of the Northeastern Seaboard. I gave old axes new edges. And I gathered friends and we took to the hills of my property in the Catskill Mountains. I met with US Forest Service trail workers, burgeoning axe forges in Maine and North Carolina, makers of premium synthetic diamond stones, third-generation lumbermen, and lifelong axe collectors. I watched as engineers snapped axes under pressure tests in a pristine lab in Upstate New York.

I took to the hills of the Catskills

I dug deep into my notes of past adventures, too. I recalled the night I slept at the Gränsfors Bruks axe factory in Sweden after a meal of *surströmming* (rancid herring) with owner Gabriel Branby. I revisited the first axe restoration class that my friend and former colleague Nick Zdon and I taught at the Best Made workshop at 368 Broadway: It was just Nick and myself, three young dudes, and a tiny woman my mother's age who turned out to be one of the most ardent axe enthusiasts I've ever met.

From Sweden to Broadway

When I started Best Made I saw the axe beginning to disappear from hardware store shelves, and from conversations and culture in general. Ten years have passed and I am happy to say I see a lot more axes out there: on social media, in stores, and for sale online. I hope Best Made played a part in that. When I started Best Made there were only three axe forges in the United States. Now I count five. (I know this sounds small, but establishing a forge in the twenty-first century is no small task.) But even with the renewed interest in the tool, books and other resources about it are scarce. This book is designed, like a good axe, as both tool and inspiration—I hope that it will serve an invaluable role in the lives of beginners and experts alike.

William Perkins,

B L A C K S M I T H,

Makes and Sells at his *SHOP* in Water ſtreet, next
Door to the Corner of South ſtreet, Philadelphia,
the following Articles—And has now for
SALE, by him.
A QUANTITY of the beſt KIND of

WOOD, or falling axes, broad

axes, adzes, carpenters malls, hatchets of
different kinds, ditching or banking ſhovels, weeding
or corn hoes, grubbing hoes, tucking hoes, chiſſels,
plain irons, 10d. 12d and 20d. nails, hooks and hin-
ges, and many other kinds of ſmith's work, too tedious to mention.   5 faw

Fiſhing Tackle.

EDWARD POLE,

At his long Eſtabliſhed

A Little History

The axe is the oldest tool known to humankind, and it's
the result of thousands and thousands of years of axe makers subjecting the tool to one of the most powerful forces
of change that humans can commit: the act of tinkering.
From the very beginning, human beings have sought to
make the axe sharper, more balanced, and more efficient.
They've strived to make a perfectly simple thing even more
perfect. Here's the history that shaped the axe you hold in   The unfolding
your hands, admittedly with a bias toward the nineteenth   continuum
century, which saw an explosion in both making the axe
and marketing it. I hope that reading it will place you along
a continuum that began hundreds of thousands of years ago,
and right in the middle of a story that's still unfolding today.

The very earliest iterations of axes were created around
2 million years ago from stones by early hominids, who
chipped off bits of material to create a jagged cutting edge.
The two-sided stone hand axe came about around 600,000   From stone
BCE. Axe knowledge spread quickly through the civilized   to steel
world: Some archaeologists estimate that about half of
the world's populated areas at that point were using hand
axes. Remarkably, they were fairly standardized: Material
aside, a hand adze from that period found in modern-day
England is indistinguishable from one found in East Africa.

The first helved axes probably came about in what is now northern Europe during the Middle Stone Age (10,000 to 8000 BCE), when a cutting edge was sharpened onto the base of a reindeer antler. By 7500 BCE, northern forest people—in the equivalent of modern-day Denmark—created an axe-like tool by joining a chipped flint celt to an antler helve. This created a better cutting edge than the antlers themselves did, which could be used to cut down small trees and make dugout boats. The first helved stone axe was created just a little later, around 7000 BCE.

From there stone axes got thinner and sharper, until they became copper around 3000 BCE, then bronze, then iron. Between 500 and 200 BCE, a heavy wedge-shaped axe with an oval shaft hole started to be produced, usually by folding a rectangular piece of iron around an oval bar to form the hole (or eye). The ends of the billet were then forged together, and hammered to create a cutting edge (or "bit").

By 1000 CE, Vikings were using a double-bit broad axe that weighed about 3 pounds (1.4 kg), with a hardened or tempered cutting edge laid onto the axe's head. There was also another, narrower, Viking axe design, which was sometimes used as a weapon but more often as a felling and trimming axe. This design was used all throughout northern Europe and the British Isles until the twentieth century. The first models in the thirteenth century were still made of iron.

The first cast steel made in large amounts was in England, around 1740. By 1765, this steel was being used in axes, but mostly for the cutting edge, since it was so expensive. Until the late 1700s, no one country was any more responsible for axe innovation than any other. Until the Americans got involved.

The first helved axe

Bronze, iron, steel

These six stone axes are from the collection of my friend John Maclean. Notice that the top four have grooves in their sides: that's where the helve would have been attached. These are older axes and could be 2000-7000 BCE. The two smoother axes (bottom) may be 800-1000 BCE. It's difficult for us to date these specimens accurately, but rest assured they are prehistoric and most likely from the American Southeast.

Throughout the eighteenth century, American axes were made mostly of iron, but with a steel bit to retain a sharp cutting edge that could be resharpened a number of times. When the edge was entirely worn away, a blacksmith would simply lay a new edge of steel over it.

But as American settlers moved west, what they needed from their tools changed. The European-style axes were not well-suited to felling the huge trees of the Pacific Northwest. Axes grew massive: At one point, axe helves were up to 42 inches (107 cm) long; felling axes as heavy as 6 pounds (2.7 kg) were sold as late as 1907.

And none of them would be considered safe by today's standards. That's because for ten thousand years few axes had a counterweight, or poll—that slab of metal opposite the bit found on any axe sold today, which provides balance and stability as the axe is swung. North American black-smiths began forging an axe with a heavier poll, providing more weight behind the helve—which in turn provided better balance. This is in essence the modern single-bit American axe.

By the mid-nineteenth century, factory methodology and standardization had worked its way in: More steel was available, which resulted in better-quality axes than the iron precursors. The first big axe company was Collins Company, founded in 1826, which became the first company to make complete axes, head and helve. Many of the advances were the result of the innovations of Elisha K. Root, a Collins machinist, who invented new processes for forging, die plating and casting, pattern rolling, and punching out material. Let's take a look at the process.

Trip-hammers, which could mechanically shape steel, replaced hand-hammering and made short work of creating an axe-head, and forge shops became highly efficient. Still, they were infernal places. Sparks showered the air as foremen yelled to workers over the clanging of tools and thundering of hammers.

From there axe-heads went to the tempering shops where they were heated in coal furnaces and dropped into water, which meant nonstop hissing and great clouds of steam rising from the pool. The grinding shop was just as cacophonous. Workers applied the heads to great grinding stones weighing more than a ton. Water was used to keep each stone cool, but dust enveloped the workers regardless. They breathed it in all day long. Many died of tuberculosis.

(Today, axes are often made using giant drop forges, in which a huge hammer falls onto the steel, deforming it instantly into the shape of a die. The eye is then punched out by another machine called an upsetter. The process still requires a remarkable amount of force, but it's much faster and much safer.)

With mechanization came dozens of tool companies specializing in axes, all of which spared no expense (or romance) in marketing them. Each tool came with its own evocative name (the Wood Slasher, the Knot Chopper, the Chip Slinger) and was emblazoned with fanciful graphics.

By the 1880s, factories could be found in towns all over the Eastern United States. Fourteen of these companies formed a conglomerate called the American Axe & Tool Company. The company instantly controlled 90 percent of American axe production. A giant factory was built in Glassport, Pennsylvania. In-fighting and legal troubles soon began straining the alliance and the company was acquired by Kelly Axe Manufacturing in 1921.

The axe had been in decline for decades, however. Refinements in saw technology were the beginning of the end of the axe's popularity. On the East Coast, common practice was to fell trees using axes, and use saws to cut them into pieces afterward; eventually, loggers realized they could use axes just for the V-notch—the back cut you make when felling a tree—and use saws for everything else.

The widespread adoption of the power saw in the early twentieth century was the killing blow; its popularity skyrocketed due to its efficiency and ease of use. Although this invention required woodcutters to tote along more gear, it was obviously much faster at cutting wood.

The twentieth century marked the decline of the axe on farms, woodlots, and lumber camps, but it also marked the advent of the American National Park system, the Boy Scouts, the Girl Scouts, the automobile, and summer camps. The outdoors was a popular new playground for a flourishing class devoted to recreation, and they needed tools to keep them dry, warm, and well-nourished. David Abercrombie, Ezra Fitch, Leon Leonwood Bean, Charles F. Orvis, Eddie Bauer, and Clinton C. Filson were a few of the first entrepreneurs, inventors, and "outfitters" to answer the call. They commissioned existing axe makers to manufacture private-label axes for their illustrative catalogs, where the axe was sold as a staple among the sleeping bags, canvas tents, and cook stoves.

The twenty-first century saw a renaissance of craft and traditional making, combined with new and easier ways to sell—on Etsy, eBay, etc. Since then, new forges have been opened, old factories have been revitalized, and interest in the axe has been renewed, marking a whole new chapter in the life of this remarkable tool.

And the
resurrection

TIMELINE

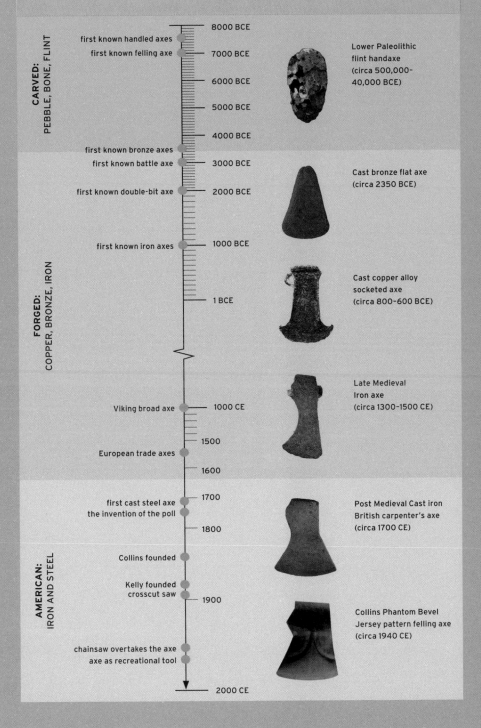

CARVED: PEBBLE, BONE, FLINT

FORGED: COPPER, BRONZE, IRON

AMERICAN: IRON AND STEEL

8000 BCE
7000 BCE
6000 BCE
5000 BCE
4000 BCE
3000 BCE
2000 BCE
1000 BCE
1 BCE
1000 CE
1500
1600
1700
1800
1900
2000 CE

first known handled axes
first known felling axe
first known bronze axes
first known battle axe
first known double-bit axe
first known iron axes
Viking broad axe
European trade axes
first cast steel axe
the invention of the poll
Collins founded
Kelly founded
crosscut saw
chainsaw overtakes the axe
axe as recreational tool

Lower Paleolithic
flint handaxe
(circa 500,000–
40,000 BCE)

Cast bronze flat axe
(circa 2350 BCE)

Cast copper alloy
socketed axe
(circa 800–600 BCE)

Late Medieval
Iron axe
(circa 1300–1500 CE)

Post Medieval Cast iron
British carpenter's axe
(circa 1700 CE)

Collins Phantom Bevel
Jersey pattern felling axe
(circa 1940 CE)

Knowing

Inside the Maryland workshop of my friend
and master woodworker Peter Dudley.

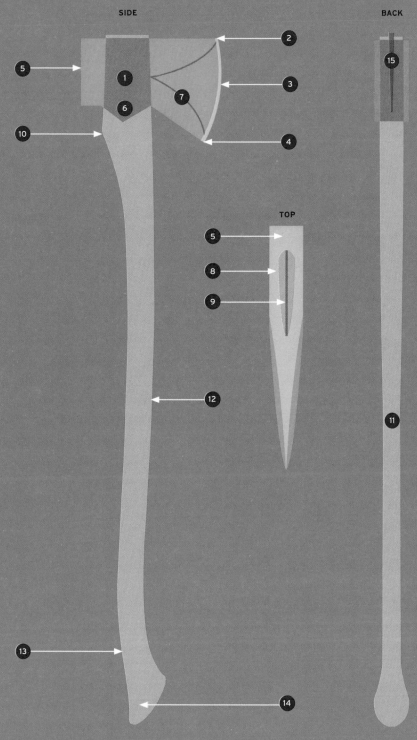

SIDE

BACK

TOP

14 / ANATOMY OF AN AXE

1. Anatomy of an Axe

THE HEAD

1. CHEEK
2. TOE
3. BIT
4. HEEL
5. POLL
6. LUG
7. BEVEL
8. EYE

In a single-bit axe, the head consists of the bit, which bites into the wood, the poll, a slab of steel opposite the bit that provides a counterweight for stability, and the eye, into which the helve is inserted. A good head is drop forged from high-carbon steel, and the bit is then hardened and tempered. The head doesn't have to be shiny and perfect—but the bit should always be sharp.

THE HELVE

9. WEDGE
10. SHOULDER
11. BACK
12. BELLY
13. GRIP
14. END GRIP
15. KERF SLOT

Often called the helve or haft, the best helves are made of straight-grained, knot-free American hickory (see page 74); they come in various shapes and sizes, but always with some sort of knob at the end to maximize the grip. Some contemporary axes have plastic or fiberglass composite helves that have few redeeming qualities (see page 76). The helve should not feel too thick or too long in your hands. It should instill in you just as much confidence as the head.

1. CHEEK   The cheek, or face, is the side of the axe-head; this is where a label or maker's mark will usually go.

2. TOE   The toe is the upper corner of the cutting edge (farthest from the user).

3. BIT   The bit is the sharpened cutting edge of the head. In early axes, it was made of steel and inlaid or overlaid on the rest of the axe-head, which was iron; in more modern axes, the entire head is formed from one piece of steel, and the bit is the sharpened portion. A double-bit axe, naturally, has two bits.

4. HEEL   The heel is the lower corner of the cutting edge (closest to the user).

5. POLL   The poll, which is distinctive to American-style single-bit axes, is a short, blunt section opposite the bit. It serves as a counterweight.

6. LUG   The lug, or ear, is a triangular or rounded extension forged into some axe-heads that offers more contact between the head and helve.

7. BEVEL   The bevel is not commonly found on most axes. It is an angular surface created along the side edges of an axe-head; bevels are intended to reduce binding (sticking within the wood).

8. EYE   The eye is an opening that extends all the way through the axe-head so that the helve (and wedge) can be inserted into it. When wedged, the helve "expands"—reducing the possibility that the head will, well, fly off. In a single-bit axe, the shape of the eye and helve end (from a birds-eye view) is usually tapered—or teardrop-shaped—with the narrow end closest to the cutting edge, allowing for better alignment and wedging.

The wedge is inserted through the top of the eye (more specifically, it's inserted into the kerf slot, a slit that has been cut into the helve with a saw), and spreads the top portion of the helve within the eye so that it fits snugly.

9. WEDGE

The shoulder of the helve is located immediately underneath the axe-head; the axe-head essentially rests on the shoulder. (The straight helve likewise has a shoulder—on a double-bitted axe, each side will have its own shoulder.)

10. SHOULDER

The back curves back toward you.

11. BACK

The belly curves forward slightly, away from you, and is where your hand will slide on the upswing.

12. BELLY

The grip is below the back and is the portion of the helve where your motionless hand will remain and where the other hand arrives at on the downswing.

13. GRIP

The end grip, or knob, is the enlarged portion at the end of an axe helve. It decreases the possibility of the helve slipping through your hands. The lower corner (which faces out toward the bit) is called the heel and the upper corner (facing the chopper and the poll) is called the toe. The end grip contributes a lot to the overall look of the axe. Most end grips are shaped into a "fawn's foot" that is then cut off flat at the end.

14. END GRIP

15. KERF SLOT

The kerf slot is the thin channel cut into the top of the helve where the wedge is inserted.

2. Axe-onomy

An axe requires a coat of linseed oil every once in a while, but it has no moving parts or batteries, and no upgrades are necessary. Depending on your demands, you may be able to get by with just one good axe. But your job is to match the task before you with the right axe—or at least know the ways in which your axe is over- or under-powered for the job—and to use that axe in the safest way possible.

The axe styles I cover in this chapter are more or less all the axes most anyone will ever need. There are, of course, nuances to each of those styles, to which I could devote an entire book. Keep it simple, and choose the axe that feels best to you. For instance, most people use a maul to split wood, but I prefer something lighter, and find that a big American felling axe will do the trick just fine. I could limb a tree with that same American felling axe, but I prefer to use my smaller Scandinavian forest axe. As your experience grows, you'll form your own opinions on which axe is right for you.

Keep it simple and choose the axe that feels best

THE FELLING AXE

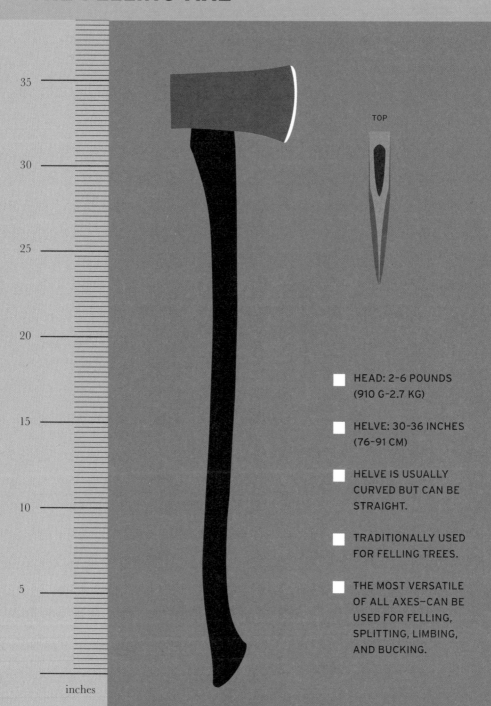

TOP

HEAD: 2-6 POUNDS
(910 G-2.7 KG)

HELVE: 30-36 INCHES
(76-91 CM)

HELVE IS USUALLY
CURVED BUT CAN BE
STRAIGHT.

TRADITIONALLY USED
FOR FELLING TREES.

THE MOST VERSATILE
OF ALL AXES—CAN BE
USED FOR FELLING,
SPLITTING, LIMBING,
AND BUCKING.

35

30

25

20

15

10

5

inches

SPECIFICATIONS:

Look for a 3- to 6-pound (1.4- to 2.7-kg) forged-steel head with a broad cheek. The helve (preferably curved) should be straight-grained unvarnished American hickory, 30- to 36-inches (76- to 91-cm) long. If it comes varnished, just sand it back, then give the shaft and the end grains—top and bottom—a coat of boiled linseed oil. You'll never forget the first time you pick one up—such a density of weight at the end of such a narrow helve. Properly matched to the user, there's nothing in the world more satisfying to swing. If it feels too heavy, simply find yourself a lighter axe.

BACKGROUND:

The felling axe as we know it was developed to meet the needs of loggers and frontiersmen on the East Coast in the eighteenth and nineteenth centuries. More efficient in design than anything that had come before, felling axe-head patterns migrated to other countries almost immediately. Most anyone felling a tree in Australia in the late nineteenth century, for instance, used a heavy-duty American-style felling axe to do the job.

USE:

It's the axe for everything. If you could have only one axe, it would probably be some form of felling axe. The bit is thin and sharp and designed to cut through wood fibers at a depth of 2 to 3 inches (5–7.5 cm). While it's not explicitly designed as a splitter, a felling axe can split soft woods, such as birch, pine, spruce, and poplar, like butter—and can handle most hardwoods, too. I frequently pick one up for this task. It can take that firewood down to kindling as well as it can limb and buck a fallen tree. No axe is more versatile, efficient, and, let's face it, beautiful.

THE DOUBLE-BIT

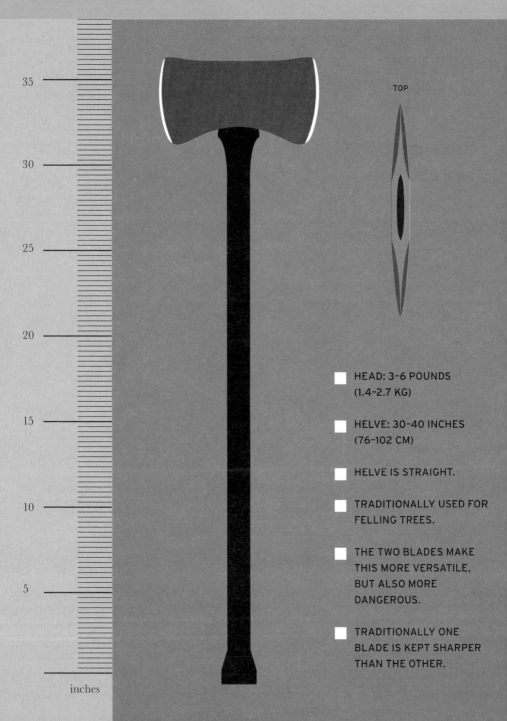

TOP

- HEAD: 3-6 POUNDS (1.4-2.7 KG)

- HELVE: 30-40 INCHES (76-102 CM)

- HELVE IS STRAIGHT.

- TRADITIONALLY USED FOR FELLING TREES.

- THE TWO BLADES MAKE THIS MORE VERSATILE, BUT ALSO MORE DANGEROUS.

- TRADITIONALLY ONE BLADE IS KEPT SHARPER THAN THE OTHER.

35
30
25
20
15
10
5

inches

SPECIFICATIONS:

The head is, naturally, much longer than a single-bit axe—by 2 to 4 inches (5–10 cm), although it won't feel much heavier in the hand. The helves are (almost) always straight.

BACKGROUND:

The double-bit was the preferred axe of American lumberjacks, especially in the Pacific Northwest, who would be in the field for extended periods. When one bit became dull, they could simply turn it over and use the other one. Two blades. One axe. Smart. Yet when double-bits began to be marketed to American loggers in the nineteenth century, many were skeptical about their safety. After all, it would seem you could stab yourself in the back with one. It didn't take long for choppers to get used to it, however, and throughout the American axe's golden age in the last half of the nineteenth century, the double-bit was considered the superior axe. I have occasionally heard double-bits affectionately referred to as "twybles" or "twibbles," but this confuses it with a small woodworking tool officially known by the same name.

USE:

Woodsmen often kept one bit extra sharp for felling, and one a little thicker for limbing—you don't want to use a thin blade for limbing because the knots you'll invariably encounter will chip it. The double-bit is a dangerous axe. In addition to impaling yourself with a blade as you begin your swing, there's a chance of stepping on or falling into one that has been carelessly driven into a log after use, as you would a single-bit axe. In *Keeping Warm With an Ax* (now sold as *The Ax Book*), Dudley Cook puts it this way: "[The double-bit's] honed edge gliding nakedly up into nothingness has an inconspicuous quality that is the essence of treachery. You can bumble into it even though you were the one who drove it into the log." Still, there is one built-in safety feature: The blade you're not using as a bit acts as a counterweight to the one you are, reducing wobble and aiding in accuracy.

THE MAUL

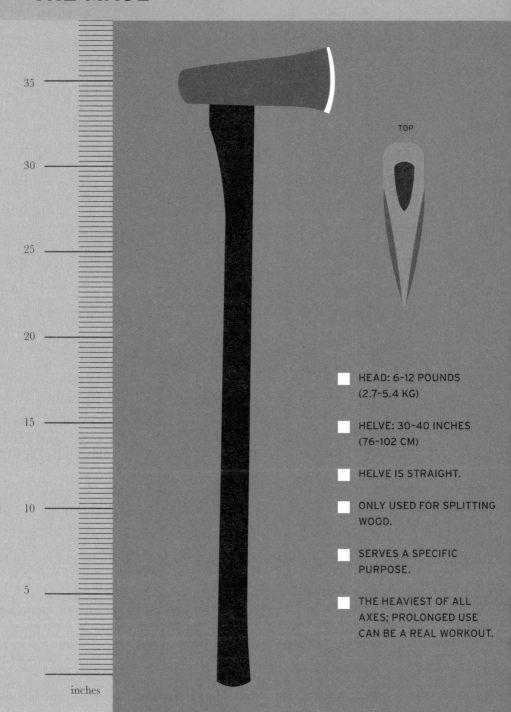

TOP

35

30

25

20

15

10

5

inches

- HEAD: 6-12 POUNDS (2.7-5.4 KG)

- HELVE: 30-40 INCHES (76-102 CM)

- HELVE IS STRAIGHT.

- ONLY USED FOR SPLITTING WOOD.

- SERVES A SPECIFIC PURPOSE.

- THE HEAVIEST OF ALL AXES; PROLONGED USE CAN BE A REAL WORKOUT.

SPECIFICATIONS:

A maul's head typically weighs 6 to 12 pounds (2.7–5.5 kg)—8 pounds (3.6 kg) is perfect—and is quite wide when viewed from above, although narrow when viewed from the side. Its helve is straight and long—at least 36 inches (90 cm). Some maul heads look quite elaborate, with indentations and raised cheeks or "wings" (built-in wedges, really) and even spring-loaded levers.

BACKGROUND:

The first splitting mauls were sold as "mauling axes" in the late nineteenth century. Before that, choppers used wedges or lighter-weight splitting axes to split wood. Ever since the form was introduced, the sheer power of the maul has captured the imagination of edge-tool marketers. One early-twentieth-century maul came with recessed levers that were activated when the maul bit into the log, pushing apart the wood. In the late twentieth century, mauls started to be sold in absurdly heavy weights: There's a 14-pounder (6.4 kg) with a triangular head being sold today. (In testing, it doesn't fare any better than a traditional lighter maul.)

USE:

Mauls are wide and heavy and meant to bite only an inch or two into the wood. After the bite, the maul aggressively rips and shatters wood fibers as it moves down through the material. You use a maul to split whole logs into halves or quarters—although experienced choppers can take a log down to kindling with this brutal tool. A maul isn't meant to be used over and over like a splitting axe or felling axe. Its weight will tire you out real quick. A maul's poll should be heat-treated so that it can be used to pound splitting wedges.

THE HATCHET

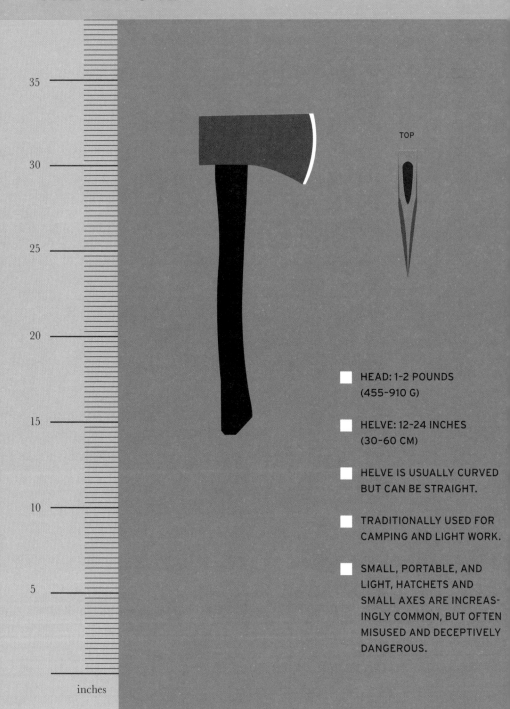

TOP

- HEAD: 1-2 POUNDS (455-910 G)

- HELVE: 12-24 INCHES (30-60 CM)

- HELVE IS USUALLY CURVED BUT CAN BE STRAIGHT.

- TRADITIONALLY USED FOR CAMPING AND LIGHT WORK.

- SMALL, PORTABLE, AND LIGHT, HATCHETS AND SMALL AXES ARE INCREASINGLY COMMON, BUT OFTEN MISUSED AND DECEPTIVELY DANGEROUS.

35

30

25

20

15

10

5

inches

SPECIFICATIONS:

The hatchet is a small, light axe intended for single-hand use. Look for a head weighing 1 to 2 pounds (455–910 g) and a curved helve of 12 to 24 inches (30–60 cm).

BACKGROUND:

Early hatchet history is confounding. There's not much evidence of hatchets in Europe. There was a great market for "trade axes" in the New World, used to barter for pelts and other valuables. I associate the rise of the hatchet with the Boy Scouts, who have made it into one of the most iconic tools of the outdoors, most often fastened to a scout's belt. Hatchets are an increasingly required piece of kit for the growing schools of survivalists and bushcrafters, and from this demand has sprung many dramatic and elaborate design upgrades, most of which are gimmicks.

USE:

It's perfect for travel, for splitting kindling, and, when very sharp, for pruning. Only the softest of woods will give way to a hatchet, so it's not to be used for splitting firewood. Be warned that the diminutive hatchet is an extremely dangerous tool. I consider it even more dangerous than the double-bit axe. Take a big soup ladle in your hand and slowly describe an arc with it, starting over your head and bringing it down in front of you with your arm outstretched. Note where the tip ends up: Not in the ground like a felling axe would, but probably in your kneecap or your femur. Spread your legs wide or, if in doubt, kneel when using a hatchet.

OTHER STYLES

BROAD AXE (1):

The broad axe looks like a kind of battle axe, with its huge face, but is in fact meant for the relatively genteel task of squaring logs to create structural timber.

PULASKI (2):

The Pulaski is a kind of double-bitted axe, with one bit being a "normal" axe blade and the other a "crossbit," which looks turned on its side. That blade is used to clear vegetation and dig shallow trenches for firebreaks. You'll sometimes see axes that look like Pulaskis but are actually "undercutter" or "chain saw" axes, in which that crossbit (narrower than the Pulaski's) is used to pick out the wedge of wood made by a saw when felling a tree.

HUDSON BAY (3):

A descendant of the early axes used for trade with Native Americans, the Hudson Bay axe has historically meant different things, but has come to be known as a medium-sized axe with a tomahawk-like 2-pound (910-g) triangular head with a deeply recessed beard. It's as handsome as an axe gets, but it is delicate: With such limited surface contact between wood and steel, it's easy to overtorque the helve and loosen the head.

CARVING AXE:

The carving axe is a small tool with a tall, curved face and a single straight bevel specifically designed for woodworking.

CAMP AXE:

The camp axe is a lightweight utility axe meant to be easily packed away for transporting. It's sometimes marketed as a "Boy's Axe."

CRUISER AXE:

One of my favorite styles is the cruiser axe, a small, light double-bit, historically used by "cruisers," lumberjacks who would survey stands of timber. They'd use this axe to mark "blazes" on trees. This is a rare type.

HEAD PATTERNS

During the nineteenth century, more than three hundred patterns were being marketed, evidence of the extreme competition and gimmickry that defined the American axe industry's first hundred years or so. By the twentieth century, there were around fifty. Look at the patterns on the following pages: These are double-bit and felling axe patterns and thus appear quite uniform (although, of course, any variation at all in the dimensions of the head can have great effect on the quality of the chopping). In fact, there are a couple of patterns, like, say the North Carolina and the Kentucky, that seem indistinguishable. Now, if you were to expand this discussion to all the axe-head patterns ever forged, you'd see more variation than you could imagine: broad axes so broad they seemed winged (in fact, there was once a Goosewing pattern), ice axes with precariously long bits, or the outlandish pancake-like turf axe forged with a completely circular bit. And then add to that axes used for carpentry and boat building (shingle axes, posthole axes, mast-maker's axes, mortising axes), axes used for building barrels (cooper's axes), and even axes specifically designed to gash the trunks of conifers to harvest turpentine (you guessed it: the turpentine axe).

DIMENSIONS AND WEIGHTS

SINGLE-BIT AXES

Head weight varies between 2½ pounds (1.2 kg) and 8 pounds (3.6 kg), although weights between 3 pounds (1.4 kg) and 6 pounds (2.7 kg) are the most common. Doesn't sound like a lot, right? But when most of that weight is concentrated at the end of a thin, light piece of hickory, it sure *feels* like a lot.

DOUBLE-BIT AXES

Head weights vary between 3 pounds (1.4 kg) and 6 pounds (2.7 kg); weights between 3½ pounds (1.6 kg) and 4½ pounds (2 kg) are the most common. Some Pacific Northwest felling axe-heads were as long as 15 inches (38 cm) and hung on helves as long as 42 to 48 inches (107–122 cm).

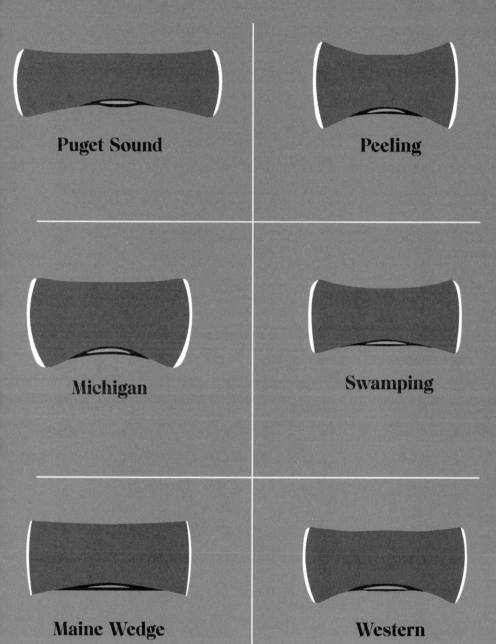

Puget Sound

Peeling

Michigan

Swamping

Maine Wedge

Western

Maine Wedge

Michigan

Dayton

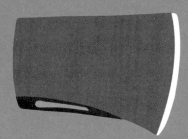

Connecticut

Narrow Wisconsin

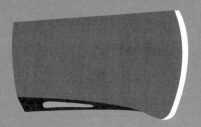

New England

Long Island

Delaware

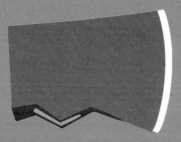

Baltimore Jersey

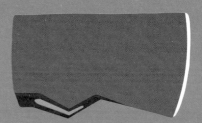

Kentucky

SINGLE-BIT
WITH LUG

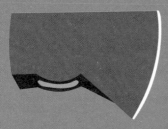

Rockaway

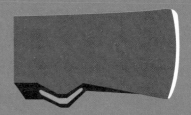

North Carolina

Double-Bit

Swell Knob

Scroll Knob

Fawn's Foot

Axes are by no means a "one size fits all"–type tool. Spend time making sure your axe fits. The first misconception is usually that bigger is better, and this couldn't be further from the truth. Some of the best and most experienced choppers I know use deceptively small axes with conspicuously thin helves. One method of sizing your axe was taught to me by Harry Prouty, a third-generation logger from New Hampshire. Pick up the axe and place the butt of the helve in your armpit. With your arm outstretched, you should be able to wrap your fingers around the top of the bit and hold the axe comfortably. Your first axe may not be the perfect fit, but keep trying other axes, always trying to match your axe with your intended use, stamina, and body dimensions. Other than "Harry's Rule," your next best bet is to just use an axe, because the more you use it the more you'll know if it fits. If it's a natural fit, then the axe will swing most intuitively, it will be a joy to use, and it's the one you'll reach for most often.

HELVE PATTERNS

Visit some American farms and you can still see the silhouettes of axe-helve patterns penciled onto the sides of barn doors and drive shed walls. There was once a time when axes were sold without helves. It was the responsibility of the user to source, shape, and hang the helve for a new axe. If you buy a new axe today, it will most likely come with a helve, but that doesn't mean you can't get out your draw knife or orbital sander and give it some shape.

President Theodore Roosevelt with axe in hand, ca. 1905.
From the Library of Congress

The President's Axe

We tend to think of Abraham Lincoln as the president most associated with an axe. And it's true that until his early twenties he used such a tool daily as a rail-splitter and woodsman—he famously possessed a 10-pound (4.5 kg) broad axe. But Lincoln's use of the axe ended when he became a politician. It's as if Theodore Roosevelt's *began* when he started running for office.

As vice president, Roosevelt once wrote to his friend William Howard Taft during a long congressional adjournment, "I am rather ashamed to say . . . [I do] nothing but ride and row with Mrs. Roosevelt, and walk and play with the children; chop trees in the afternoon and read books by a wood fire in the evening."

Despite such self-deprecation, he wasn't ashamed at all. He couldn't have been prouder with how he spent his time. I doubt it was a coincidence that when a *Ladies' Home Journal* reporter came calling on Roosevelt a few months after he became governor of New York, he was hacking away at a hickory. The reporter wrote:

> Long before I reached the spot whence the blows of the axe were booming out into the snowy air I knew that the new governor was unquestionably cutting straight and true. The deep-chested, full-toned "Hep!" that accompanied each blow of the axe was quite enough of a certificate for that.

Almost every magazine profile of Roosevelt included a scene involving an axe. "The President cleans up his own woods, liking nothing better than to see the chips fly and to hear the sharp ring of the ax," extolled *McClure's* in 1906.

John King of King Axe presented Roosevelt with a 2-pound (0.910 kg) axe with a 14-inch (36 cm) hunting knife embedded in the helve on a campaign trip in Waterville, Maine, in 1902. As the train pulled away from the station, Roosevelt waved the small axe over his head for the crowd to see. It was a symbol to him. (It was a powerful symbol for King, too. The President's Hunting Axe soon went into mass production, and you can find examples at auction today.)

The axe was an extension of his image, an emblem of his presidency— during which, the national park system was founded—and a powerful and useful symbol for the labor and skill that was literally building his country.

That same attitude drives our current interest in the axe. Of course, we could employ a power tool to do a quicker job of things. But we use an axe because it does the job quietly and efficiently and because it connects us to the outdoors in a way a chain saw or wood splitter can't. The axe sharply rang true in Roosevelt's time in the same way it sharply rings true today.

3. What Makes a Good Axe?

The best axes I have ever swung are efficient tools that elicit unbridled joy. On the joy meter of 1 to 10 (10 being the most joyful), you should swing an axe that prompts a 9 or 10. Don't settle for anything less than 8, and don't assume that an expensive axe will mean happiness. Some of the best axes I have ever swung have cost me less than $30 (all in). But we'll get to that later in the book. Time spent chopping wood and being outside is sacred, so respect that time by investing in an axe that's a joy to use.

A good axe, when held in my hands, imbues me with a sense of confidence, calm, and delight. I think it's important to surround ourselves with things that have this effect, the things we love, things with meaning, and things that we want to use. At the end of the day, the best things in my life just feel right; they belong.

A good axe
imbues
confidence

It's easier to know when you're holding a bad axe, of course. You'll have no problem spotting a chipped blade, or a splintered helve. It pays to sweat the small stuff. If you don't see any obvious problems but something still feels off, then it probably is off. When it comes to buying an axe, ask a lot of questions, and if those answers don't add up, or worse yet, if they can't even be answered, then keep looking. A good axe may not always be easy to find, but the good news is they are out there (the fun part is finding them).

It's easier to forge a dull axe than a sharp one, because forging a dull axe requires less skill, and less time. If an axe is sharp it probably (but not always) means the maker has gone the distance, not just on the blade, but hopefully the rest of the tool. At face value, an axe forged in a lesser-quality steel with the right blade profile and bit geometry is superior to an axe forged in a higher-quality steel with a lesser profile and geometry. But keep in mind you can almost always improve upon a bad bit (see page 124).

Keep an eye on the grain orientation along the helve—it speaks volumes of the maker. If you hold the axe up with the butt end pointed toward you, you should be able to see that the grain orientation runs parallel to (or in line with) the orientation of the blade. The manufacturer obviously doesn't make the wood grain, but it is responsible for selecting the right wood, hanging the axe, and making sure it is all in alignment. Just keep in mind there are always exceptions to the rules: Grain orientation mainly applies to larger axes (28 inches/71 cm plus), and sometimes the right orientation could just be a lucky mistake on the part of the maker.

Grain orientation

An axe is just a piece of steel mounted to a slim stick of wood. Simple, right? Well, there's a mind-boggling array of steps that go into making an axe. The bit may be sharp, the grain in alignment, but if there are gaps in the eye, or if the head is not seated correctly on the helve, then you will have problems. Don't let the relative simplicity of an axe fool you: Every excruciating detail must be considered when making an axe, and it's your job to evaluate those when buying.

Every detail must be considered

As for materials, the helve of any axe will most likely be made of American Appalachian hickory, or plastic. Back to the happiness meter scale of 1 to 10, a plastic-helved axe is less than a 1: It will be painful to swing, it will be impossible to maintain, and it should be avoided at all costs (see page 76). Don't be alarmed if you happen upon a helve made of ash—that's what they make baseball bats out of, and it's a perfectly acceptable, albeit an uncommon and dated, choice.

Hickory, ash, but never plastic

Steel by its nature won't reveal itself as easily as wood. For new axes, the maker or retailer should offer the origin and type of steel in question. What you do with that information is up to you, but at least you have it. Most axes I know are forged in either an American or European steel, and I would challenge anyone (without a mass spectrometer) to tell me the difference. European steels are thought to be slightly softer, and their profiles are designed for softer European woods, but the European axes I swing work just great on my dense American sugar maple or black cherry. For old axes, there may be a mark that identifies the origin of the axe (though not necessarily the steel). Be it new or old, you want an axe that is forged with high-carbon steel (see page 50). There is very little information about the composition of steel in old axes (pre-1950, say), but some say that it's superior to steel in newer axes because blacksmiths and forging techniques were better back then, and the steel was purer (i.e., not recycled). This "old is better than new" theory is convenient, and to my knowledge remains anecdotal.

Look for carbon steel

It's no coincidence that the best axes I've ever owned happen to be a 10 on the happiness meter and a 10 on the joy-to-look-at meter. Like sharpness or grain orientation, you should consider the proportions, the silhouette, the shape, and the lines of an axe. How much thought has the maker put into this tool overall? Does the axe speak to you? If so, what does it say, and is that a good enough reason to part with your hard-earned money? Axes make bad impulse buys, so take your time making that final decision.

WHAT MAKES A GOOD AXE?

STRAIGHT GRAIN ORIENTATION	For axes 28 inches (71 cm) or longer, make sure the wood grain runs parallel to the blade (see opposite).
A SHARP BIT	A keen edge means efficiency and safety, and it's a good indication that the maker has invested the necessary time into the tool.
A SECURE HEAD	Check for even the slightest looseness in the connection between the head and the helve, and gaps in the eye.
QUALITY MATERIALS	Carbon steel and American straight grain hickory are the starting point.
A SKILLED MAKER	There's nothing worse than an anonymous axe. A maker's mark should be a badge of honor. Know who made your axe.
PERFECT ALIGNMENT	When you are sighting down the length of the axe, the line of the head should not deviate a hair's breadth from the line of the helve.
BALANCE	This can be assessed only by holding the axe with both hands: Does it feel too top-heavy?
FIT AND FINISH	In a tool as seemingly simple as an axe, every detail matters. Scrutinize the tool, and if you're not happy ask for options.

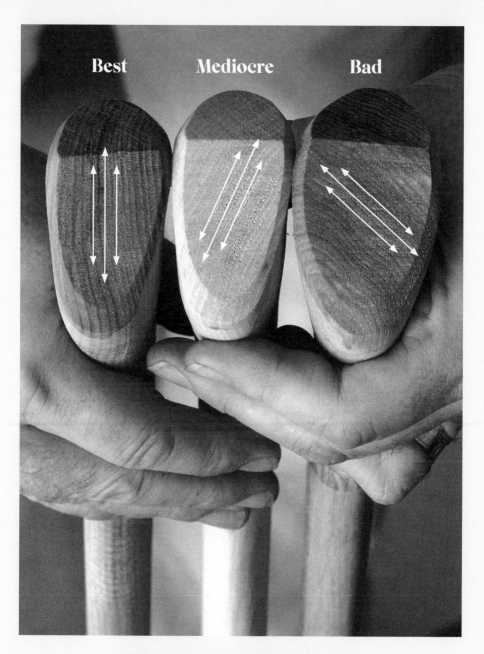

GRAIN ORIENTATION

When considering an axe that is around 28 inches (71 cm) or longer, sight down the helve vertically: The end grain (seen best at the butt of the helve) should be in line with the plane of the axe-head, not perpendicular to the central axis.

4. Dynamics of the Head

In every restoration class we taught at Best Made, I could sense a little panic during the seconds before the students' files made first contact with the blades—that's the almost universal fear of ruining the perfectly machined edge of your first axe. I get it: An axe-head can be a pretty intimidating thing. So it pays to know the workings of the head inside and out—to know how a sharp, cold, heavy piece of steel becomes a tool that's got character and, dare I say, some soul.

Because the axe is a tool carried and swung by a human being, the weight of the head is really important. While the splitting mauls range from 6 to 12 pounds (2.7–5.4 kg), this is a considerable amount of weight to be picked up and dropped the hundreds of times it takes to split several cords of wood. Although a 10-pound (4.5 kg) felling axe would generate an incredible amount of chopping power, it is unreasonable to expect the average mortal to wield such an axe with any accuracy or efficacy. The typical felling or all-purpose axe clocks in between 2½ and 4 pounds (1.2 and 1.8 kg), which is a range that allows a variety of swings, tireless chopping, and ample striking power. A heavier axe will not make you better at chopping wood; the best axe is the one that fits you the best (see page 35).

The right weight

As the axe-head strikes its target, the edge is the first part to interact. If it's sharp, the edge will bite into the wood fibers and guide the axe's energy directly into the intended cut. If the edge is dull, the axe might glance off the target wood, causing the cheek to smack the surface and the reverberations to resonate down the helve. While a sharp edge slices, a dull edge tears, and that's the difference between where you want the axe to go, and where the wood wants it to go. A glancing blow may lead the axe directly into the dirt, or worse yet, your boot. It can't be said enough: A sharp blade is a safe blade, and the work of a sharp edge is always more efficient. A microscopic view of a steel edge reveals that not even a highly honed edge is entirely uniform, but instead consists of a series of micro-serrations (see page 48). There are flat spots at the base of these serrations, which are minimized by the sharpening process. As a blade dulls, these flat spots become larger and begin to tear fibers rather than slice through them. Thus the finer an edge is honed, the longer it will last and better it will perform, yet every chopper has a limit to how obsessively they want to sharpen their tool.

Your axe must be sharp

The goal of an edge is to have a width approaching zero at its finest point, with a profile that fits the blade's intended purpose. There's no blade more wicked-sharp than the concave straight razor, which surpasses all in slicing ability. However, if we designed an axe with the thinness of the straight razor, it would crumble on the first strike. To create an edge with a keen point and plenty of meat to endure a life of repetitive shock, the axe is ground with a convex bevel, which leaves more steel at the area of impact than other bevels and still allows a fine edge. Like most knives, the bits of axes are commonly ground to a V, also known as a double bevel. Specialized carving axes, like some Japanese chef's knives, are ground to a single bevel (see opposite). Although the sharpening of a convex profile is more difficult to master, it's worth the practice

But not too sharp

BIT GEOMETRY

OTHER BIT PROFILES

DOUBLE BEVEL

Strong, but less efficient than convex.

SINGLE BEVEL

Common on specialized carving axes.

CONCAVE

Weak and unsuitable. Avoid.

THE CONVEX PROFILE

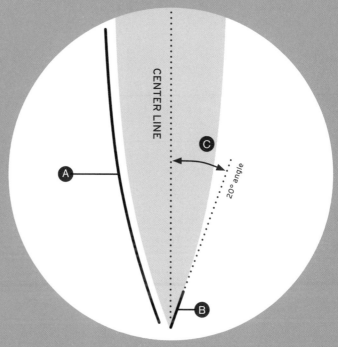

CENTER LINE

20° angle

Ⓐ

Ⓑ

Ⓒ

MAGNIFIED CROSS SECTION OF THE CONVEX BIT

This profile offers the best balance between strength and efficiency. It's the profile I recommend for almost all axes. The primary feature of the convex profile is the lack of separate primary and secondary bevels. Like a bullet, the bevel angle of the convex profile gradually changes as the bit transitions seamlessly into the cheek of the axe-head. The overall thinness of the cutting edge is determined by measuring the angle of the first quarter inch of the bit. For general purpose axes, something for splitting and felling, 20° is a good angle. Axes dedicated to felling can be a bit thinner, and those dedicated to splitting can be a bit thicker.

Ⓐ The transition from the bit to the cheek of the axe-head should be smooth and gently curving.

Ⓑ Due to the lack of wide flat bevels, angles should be measured at the last quarter inch of the bit

Ⓒ 20° is a good all-purpose angle. Felling axes may be 15°. Splitting axes may be up to 30°.

BITS MAGNIFIED 100x

Dull

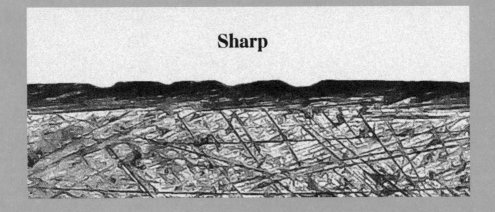

Sharp

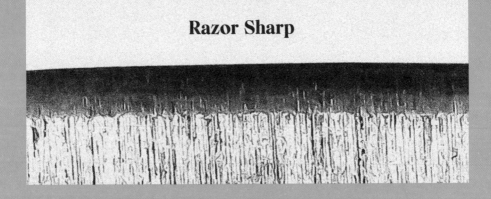

Razor Sharp

to maintain. A convex bevel will stand up to the rigors of chopping and keep it's edge longer than other profiles.

As the bit buries deeper into its target, the cheeks come into contact with the interior of the wood. The width and angle of the cheeks determine the shape of the wedge that's doing work on the wood. A thinner wedge will cut deeper, yet will provide less outward force to encourage the splitting apart of the wood fibers. A wider wedge will generate much more outward force, yet requires more input energy to transfer into outward force. This illustrates the profile difference between a felling axe and a splitting maul. A felling axe is thinner, designed to cut deeply into green wood, throw chips, and maintain a usable weight of 3 to 4 pounds (1.4–1.8 kg). While a felling axe can easily be adapted for splitting wood, if it is overconfidently aimed at the center of a large round, it can sink in and bind, making it hard to remove. The cheeks of felling axes have slight concavity to counteract the wood's habit of squeezing back onto a buried axe, though many of us have experienced the frustration of extracting a stuck head from a stubborn log.

How the shape works

The much wider wedge of a splitting maul, which often features even wider flanges, speaks to its sole purpose of cracking apart round after round. The maul is not required to be swung horizontally, or even precisely guided, so much as lifted and dropped; the heavy head and wide bevel do the rest of the work. The maul still needs a sharp edge though.

The butt end of the single-bit axe is the poll. The poll isn't an afterthought. It contributes significantly to the overall connection between helve and steel, and is sized by the toolmaker to achieve the target weight. The poll of a felling axe is usually not hardened, and you should never use it as a hammer or hammer on it with other steel tools. This will deform the poll or the eye, which is worse, and leave scars that look terrible and ruin the axe. Good judgement allows

How the poll works

the use of axe polls for tapping in plastic (not metal) felling wedges or knocking a round into proper position. Good splitting mauls have hardened polls shaped into sledge heads that can hammer steel wedges for splitting full logs. Job-specific axes have specialized polls, such as the spike of the firefighter's Pulaski, or the milled-face hammer of the carpenter's hatchet.

Historically, double-bit axes were the choice of lumberjacks for their versatility, longevity, and accuracy, yet the double-bit does not make the best all-round utility axe. The idea behind a double-bit is that a logger making wages based on trees felled in a day had an interest in spending as much time with a keen edge and as little time sharpening as possible. The double-bit can provide two keen edges with the weight of one axe, or one keen edge and one "utility" edge that spares the fine edge from ugly work such as chopping knots or roots. The balanced weight of the symmetrical head affords greater striking accuracy, and the straight helve creates less error of deflection. While the double-bit may be a great tool for the professional, it doesn't make for the best utility axe. The double-bit is inherently more dangerous: It can't be safely carried unsheathed while cruising, and presents a hazard when lying around the woodpile.

The double-bit

Every modern axe-head is built from steel, so it's worth understanding how the material behaves. Steel is, most simply, iron with a small amount of carbon. A 0.1 percent change in carbon content dramatically changes the characteristics of the steel, thus carbon content is often the primary signifier of steel types. The American Iron and Steel Institute (AISI) classifications are as follows: low-carbon steel contains 0.05 to 0.30 percent carbon, medium-carbon contains 0.30 to 0.60 percent, and high-carbon steel contains 0.60 to 1.00 percent. Steel can contain up to 2.1 percent carbon before it becomes cast iron. Plain steels only contain iron,

The composition of steel

carbon, sulfur, and manganese. Sulfur is present in all steels as an impurity, so manganese is added to counteract the embrittling effects of sulfur. Alloy steels contain at least one additional element to adjust its characteristics to specific purposes, such as the addition of chromium and molybdenum to produce stainless steels. Axes are most commonly made from high- or medium-carbon steel, which offers the ideal balance of hardness, for taking a sharp edge, and toughness, for resisting breakage when chopping hardwood. Scandinavian axes are thought to have a higher hardness than American-made axes, because they are primarily used on European softwoods and do not require the same toughness.

Steel's hardness is the metal's ability to resist deformation, and toughness is the steel's resistance to breakage. The higher the carbon content, the harder a steel can be, and the harder the steel is, the less tough it will be. The edge on harder steel bits will last longer; however, the steel will be more brittle and chip more easily. Steel is hardened by heating it to high temperature (around 1475°F, or 800°C), then quenching it in an oil or a water bath. At this point it's as hard as glass and just as brittle. If dropped on the shop floor, it will shatter. The steel is then tempered by heating it at a lower temperature (around 750°F, or 400°C). Tempering relaxes strain in the internal crystal structure of the steel, reducing the hardness to its intended range. Failure to maintain the proper temperature for the adequate time can ruin the blade, and the heat treatment process must be repeated. Hardness of blade steel is often reported in HRC, or the Hardness Rockwell C scale. Standardized hardness scales are determined by pushing a very hard implement into the surface of the subject steel with a given weight load. The diameter or depth of the impression left in the surface of the steel is translated into a

Caring for
steel

hardness value. The HRC scale uses a conical diamond bit loaded by approximately 330 pounds (150 kg), with values of blade hardness typically seen between 40 HRC and 65 HRC. The HRB scale is used for softer steel, and the HRA scale is used to measure surface-hardened steels. Axes often clock in at the low- to mid-50s HRC, while a razor-sharp sashimi knife will have an HRC above 60.

Hardness of steel

A quality axe is edge-hardened, meaning the steel is quenched only at the edge, allowing the posterior of the head to cool slowly and soften. The resulting bit has hardened steel only where it is needed—at the edge—and softer, much tougher steel around the cheeks, eye, and poll, where toughness is an asset and the overall durability of the axe is improved. Alloying elements will improve the depth to which the edge steel is hardened, which comes into play over the lifespan of the tool. As the blade is sharpened, steel is removed, meaning there is a limit to how many times a blade can be sharpened before the hardened steel gives way to soft. This lifespan is far beyond the needs of the average user, unless the user is an aggressive sharpener or has removed several deep chips.

Alloy elements

Forging is the optimal method to form an axe-head, as the compressive force of hammering results in a tighter and more uniform grain structure. Just as the iron and carbon atoms form a crystal lattice at the smallest unit level, the crystals naturally form grains that comprise the larger solid. Axes are most commonly made by a skilled blacksmith using either open- or closed-die drop forging processes. In open-die drop forging, the hot steel is placed on a die under a large mechanical drop hammer. The blacksmith releases the hammer, and it crashes down on the hot steel, pressing it into the mold and giving it the desired shape of the axe. This is repeated through a series of dies until the axe has been forged. In one fell swoop, closed-die forging stamps

Drop forging

the axe-head out like a cookie cutter, and requires less artistry and guesswork. In both cases, the effects of forging on grain structure contribute greatly to the strength of the steel. The finer and more uniform the grain, the less strain on the atomic crystal structure there is, and the less likely the steel is to have weak points or lines of fracture. Proper heat treatment can also reduce grain size; this requires quickly heating the steel, holding it at the required temperature for the minimum time, and evenly but rapidly quenching the blade.

The chemical recipe of steel is adjusted for its intended purpose. Plain carbon steel is an economical way to produce an easily worked, easily sharpened blade that maintains an edge well. It rusts easily without care, however. Stainless steel is produced with the addition of chromium and molybdenum, which forms a relatively nonreactive surface layer that undergoes oxidation much slower than exposed carbon steel. While stainless steel is not entirely immune to rust, it requires less frequent maintenance. Other elemental additions include silicon for increased toughness, tungsten and cobalt for higher hardness, and vanadium for refined grain structure. While knife blades are made from an overwhelming list of steel acronyms and numbers, axe steels tend to be fairly simple, the most common being 1060 and 5160. This is largely due to the amount of steel required to forge an axe-head.

Carbon steel vs. stainless steel

To maximize longevity, steel should be given a light surface coat of a nonrancidifying oil such as mineral oil regularly, and especially before long-term storage. Steel blades should never be stored long term in sheaths, as leather can trap moisture against the surface of the blade. More frequent, light honing will keep the blade performing at its best, and spare you the chore of long sharpening sessions.

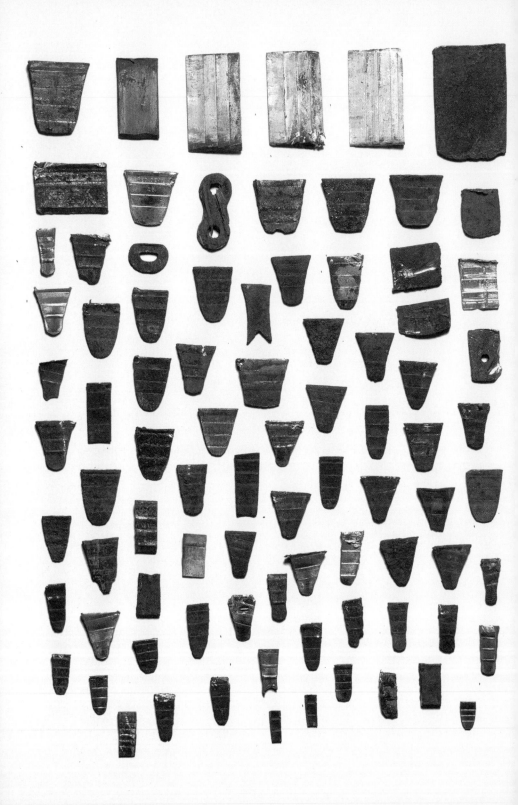

The Most Underrated Part of Your Axe

After many years swinging an axe, I've come to an important conclusion: Axes generally work best when the head stays attached to the helve. There's one critical but often overlooked piece preventing this from happening: the wedge.

A store-bought helve will arrive with the top notched so that the wedge can be driven right in once the helve is secure in the eye. The wedge squeezes the wood against the interior surfaces of the eye, making wood and steel a single unit. Any available hardwood is fine for making a wedge; the wood must be able to take a pounding without splitting apart inside the helve kerf, and must have the density to uniformly expand the flanges of the hickory helve.

A secondary steel wedge, small enough to fit laterally or at least diagonally inside the eye, is often inserted to make sure the wedge *never* comes loose. If the wooden wedge is the exact right size for the width of the kerf, and properly trimmed to length, there is enough positive traction that no secondary wedge is required, but this requires careful fitting (see page 171).

Note: The steel wedge makes it a lot harder to replace the helve, and may cause accidental splitting of the wood inside the eye.

Once the helve has been properly oiled, the wedge and top of the helve will swell even further, and once the oil has polymerized, the wood inside the eye will require drilling to separate helve from head.

5. Making

It's easy to think of everything associated with the axe as somehow "old." Ancient techniques and ancient materials combine to make an ancient tool. And so when I started this book, that was all the more reason to tell the story of someone, well, new.

The obvious choice was Liam Hoffman, presiding over his pristine forge (filled with old and new machinery) down a remote back road in the Blue Ridge Mountains of North Carolina. Liam started forging at age thirteen, while he was earning his Eagle Scout badge. His parents gave him a small brake drum forge for his fourteenth birthday. In his early twenties, with ten years already under his belt, he became the youngest winner of the History Channel's *Forged in Fire* competition show. Now Liam is at the helm of his thriving forge creating axes that people wait years for.

Liam and his team of five (including his mother Karen, who designs and crafts all the leather work) are making axes unlike anything I have ever seen, using techniques and processes unlike anything I have ever seen. This small gallery of pictures showcases both how axes have been made for centuries and how a young forger is changing the craft.

The Head There are two common ways to forge an axe: open- or closed-die. Open-die forging, practiced by Hoffman, involves shaping a piece of steel between two opposing dies that do not enclose the material. This is more free-form and allows for more control and artistry, whereby every finished axe is unique and bears the hand of the maker. The closed-die forging process is more akin to a cookie-cutter technique, better suited to mass-production whereby the axe is stamped out of dies that enclose the steel in one fell swoop. Hoffman starts with a billet of 4140 steel that is heated to 2000°F (1093°C) in a propane-fired forge. Hoffman continues to shape the axe using a series of open dies, all the while drifts are manually inserted to ensure the eye does not collapse. The maker's mark is then stamped into the hot steel. Working quickly, Hoffman himself (facing page) grasps the hot billet with tongs and passes it through a gauntlet of presses and power hammers and the axe starts to take shape (above). The art of making axes is in the dance between the delicacy of malleable steel and the brutal force of a hydraulic press. Like any great artist, Hoffman makes it look easy.

Shaping This circa 1930s Loshbough & Jordan punch press is used to shape the axe and smooth the blade. A hydraulic press is used to push a drift, or channel-making punch, through the center of the billet, which will form the eye of the axe.

Grinding is often required during the open die process to remove excess steel. Hoffman heat-treats each axe (including the poll) in a furnace, and then quenches it in a specially engineered oil that hardens the steel. Afterward, the axes go back into the furnace where they are slowly reheated, a process that "draws the temper" of the axe.

Sharpening

Fully self-sufficient, Hoffman forges
his tongs, many of his blacksmith
tools, custom dies for the forging
equipment, and almost any part
needed to run the forge. Here he uses
a custom-built jig to sharpen the bit
of a carving axe on his Wilmont
belt grinder.

The Helve

The copy lathe (above), the centerpiece of Hoffman's handle shop, sits against the back wall. Four hickory blanks are hand-selected to ensure premium grain quality and are then fixed into the lathe, where they are turned into rough handles. Using a 30-inch (76 cm) band saw, Hoffman cuts the kerf slot into the top of the handle, which will receive the wedge. Few axe makers make their own handles, let alone their own wedges. The opposite page is a detail of a Hoffman walnut wedge, with rounded ends to ensure a near-perfect fit in the expanse of the eye of the axe. Wedges are shaped to fit the axe, the handles are shaped to fit the eye, a little wood glue is applied for good measure, and then it's carefully forced together with a 12-ton arbor press, leaving a secure and lasting fit. No stone at Hoffman's is left unturned. The last step is to make the leather blade guard. Hoffman works closely with his mother Karen, a skilled artist and craftsman, who hand-cuts, sews, rivets, and finishes every Hoffman blade guard.

6. Dynamics of the Helve

When you swing an axe you grip the helve, sometimes for dear life. That slender piece of wood is your lifeline, your accelerator, your brakes, and your steering wheel for guiding a very sharp and fast-moving billet of steel. As we saw in the last chapter, the head of an axe can be a cold, mysterious, and complex thing. I think the helve is something much more familiar—warmer, more organic, tangible—in more ways than one. For all my talk of maintaining a good sharp bit, you should never lose sight of your connection to the helve. Only use an axe that you're comfortable holding, and if you have to get out your drawknife and sander to make it so, then do it.

Helve length can range from 8 inches (20 cm) on a hand hatchet (just enough to distinguish it from a Neanderthal's axe), to over 40 inches (102 cm) on the double-bit foresters' axes used to bring down the old giants of the American West. General-purpose axe helves are found in the 28- to 36-inch (70–90 cm) range, while camping axes are shorter

Helve shape and length

for portability. Helves are ideally made from hickory (see page 74 for why that is), though other durable hardwoods such as ash are occasionally used.

Most modern single-bit axes have a curved helve, while double-bit axe helves are always straight. Helves are thicker at the shoulder near the head for positive engagement between the helve and bit, and to add strength where the greatest shock occurs. Straight single-bit helves are considered by some to be more accurate than curved helves, but I think it's a matter of personal preference. Pick up an axe with a curved helve and point it instinctively toward a distant target. Now try the same with a straight-helved axe. Decide which feels more natural, and go with that.

Straight helves are sometimes less expensive, and can be more easily made from a split section of wood, if you're shaping a helve from scratch. A split-grain helve will typically be stronger than a sawn-wood helve, as the split wood cleaves along the tree's natural lines of vertical strength. Saws ignore the wood's proclivities, cutting along an arbitrary line that may not play to the wood's strength, resulting in grain run-out. The end grip at the base of the helve keeps the axe squarely in the chopper's hands, preventing the axe from flying out of your hands. Older helves sometimes have a pointed fawn's foot shape, while any modern axe will have a flat or scroll end. While the fawn's foot was an elegant touch and quick to carve, it is prone to breakage and is an annoyance during the helve-hanging process, whereas the flattened end of a scrolled end grip offers the ideal surface for a solid mallet strike.

Straight vs. curved

You would think, as many makers would have you do, that thick is better than thin when it comes to a helve, but this is a pervasive misconception. Think of your axe as a holy matrimony of wood and steel, as one entire system. Unless it's a maul, the eye of most any axe is actually quite thin, and it turns out there's no reason your helve should be any thicker. The load, or resistance, is channeled through the head, and so the success (or failure) of an axe greatly depends on the point where the head and helve meet. The more seamless the connection, the better the system. A thick helve has to be made thin right where it's crammed into the small eye of an axe, and that creates an acute transition—and a breaking point. The helve should naturally flow from the eye. Makers of most older axes knew this, and they have exceptionally slim helves.

Your axe is a system

When hanging a new helve, it's vital to select a quality specimen. The wood should have a clear, straight grain—the closer the grain is to running the entire length of the helve the better. Similarly, when sighting down the helve vertically, the end grain should be in line with the plane of the axe-head, not perpendicular to the central axis. When the grain crosses the helve laterally, or changes direction toward the edges (the run-out), this creates a line of fracture. The idea is to split the firewood, not the axe. It is harder to find a curved helve with grain that stretches the entire length of the helve, as production helves are cut on a lathe, not bent, which gives the straight helve shape a slight advantage.

Selecting a helve

Knots create weak points in the system of the axe, as the grain twists around the knot's center of high density. They form when a tree's attempt at growing a new limb fails, and the healthy trunk grows around where the limb previously connected. The limb disrupts the strong, longitudinal grain, and creates a void where the dead limb dropped off. Rather than a continuous, fluid flex, the helve now has a point of sharp deflection at the knot, improving the chance of breakage. The grain should not be too tight, either. Think of the strength difference between a bundle of small twigs and a bundle of thick limbs. Both are

The problem with knots

Helves are cut from logs of locally sourced Appalachian hickory at the Tennessee Hickory Handles factory in Loudon, Tennessee.

stronger as a bundle, but the individual limbs are less likely to break under a sharp blow than the individual twigs.

If choosing between two clear, long-grain helves, pick the helve with more meat between the grain lines. Some helves contain both heartwood and sapwood, indicated by a sharp contrast change in the wood. The lighter wood, sapwood, is the younger, "living" rings of the tree, while the darker heartwood is the older core of the tree whose vascular system is no longer functioning. The USDA Forest Products Laboratories have determined that there's no strength difference between heartwood and softwood, and a helve containing the transition between the two does not affect its longevity. Choosing a helve with heartwood, sapwood, or a combination of the two is a matter of aesthetic preference.

Heartwood vs. sapwood

Helves are made from kiln-dried wood, which continues to shrink and swell as its moisture content changes. If left untreated, a raw wood helve can shrink so drastically that the head becomes dangerously loose. To make the helve more dimensionally stable and durable, it should be coated or soaked in a drying oil. The best choice for this utility is boiled linseed oil, which is linseed oil with additives that expedite curing and hardening. You could also use tung or walnut oil, but these options will be a lot more expensive than your hardware store linseed. Raw linseed oil produces a similar result but takes longer to harden. Here's how it works: Wood soaks up the drying oil, which polymerizes and hardens as it cures, essentially "capping off" the wood fibers, preventing them from soaking up the moisture that causes them to swell and shrink. A surface layer is formed as well, improving the overall weather resistance of the helve, without being so glossy as to reduce positive grip. Helves can be oiled by either submerging the axe-head in oil and allowing the helve to naturally wick

Linseed oil

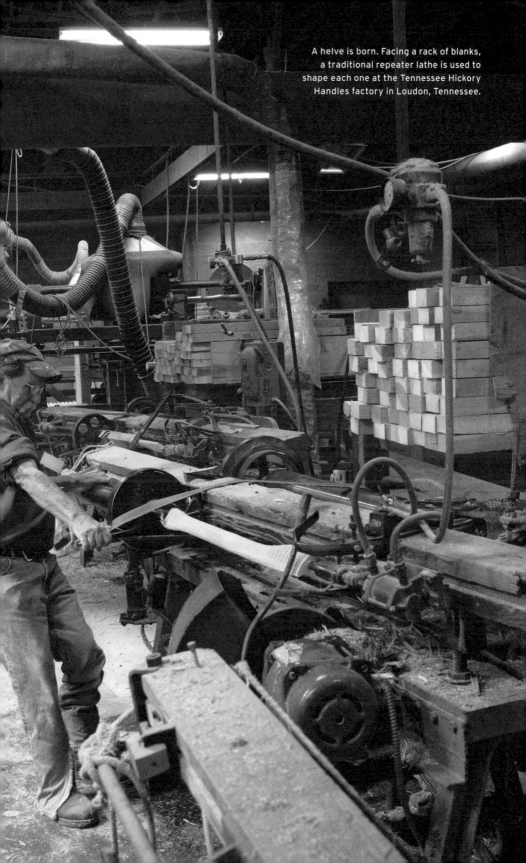

A helve is born. Facing a rack of blanks, a traditional repeater lathe is used to shape each one at the Tennessee Hickory Handles factory in Loudon, Tennessee.

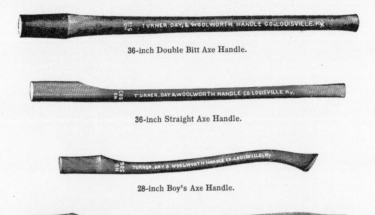

36-inch Lumberman's Axe Handle.

36-inch Double Bitt Axe Handle.

36-inch Straight Axe Handle.

28-inch Boy's Axe Handle.

36-inch Single Bitt Axe Handle.

HICKORY AXE HANDLES.
Oval, Plain Ends.

Why Hickory?

Almost every axe the world over will have an American hickory helve, most likely originating from somewhere in the Appalachian forests of Arkansas, Tennessee, or Missouri. The modulus of elasticity is a measure of how much a material resists deflection before it deforms or breaks, and is commonly used for determining structural strength of lumber. American long-grain hickory has the highest modulus of elasticity among hardwoods, and thus the greatest stiffness and strength. But remember: Wood has different strength properties depending on grain and material orientation. It's equally important that a hickory tool helve has grain that is parallel with the direction of force. Ash is perhaps the next best wood of choice for an axe helve. In Japan they might use oak, but nothing is quite as good as straight grain hickory.

the oil internally, or by applying the oil to the surface with a brush or rag in a series of coats until the wood is no longer taking up more oil. If you plan on collecting and/ or restoring a lot of axes, you might want to invest in a linseed oil tub that you can immerse your axes in (completely or just the heads) overnight. Regardless, attention should always be focused on the end grain—the top and bottom—of the helve, where the vast majority of oil absorbance occurs. And be watchful where you store your axes. Leaning an axe up against your fireplace mantel might make for a pretty picture, but common sense would dictate that prolonged heat from the fire is going to dry your helve up. As kids, we used to soak axes with loose heads in water overnight, and this was a terrible idea. Yes, the water would expand the helve in the eye of the axe, giving us some temporary relief and a somewhat secure connection, but soon enough, when it dried it would shrink the helve even further, leaving us with an even looser, and more dangerous, head. Plus some rust for good measure.

One last note about linseed oil: It's critical that you properly dispose of any rag, paper towel, etc. that has come in contact with linseed oil; left unattended they have a likely chance of spontaneously combusting (see page 206).

Common hardware store axes are typically sold with varnished helves, which is an inferior option. Varnish completely seals the wood and is highly durable, but the surface tackiness of varnish is a quick way to earn a handful of blisters, and the discoloration of the varnish that comes with age is not nearly as handsome as an oil patina. A small painted sash at the base of the helve aids in locating the axe when returning to a fallen tree in the woods, or when searching for an axe mislaid behind a woodpile.

Varnish and paint

THE WOOD VS. PLASTIC TEST

I met Jared Nelson, assistant professor in the Division of Engineering Programs, at his lab in the State University of New York at New Paltz, holding six shiny new axes in my hands. The only difference between the axes were their helves: three were plastic, and three wood (hickory). Our goal was to objectively understand the differences between wood- and plastic-handled axes. Jared and I subjected the axes to stress tests under an impressive 150-kilo Newton press (opposite). With more time and resources, we would have built custom fixtures to secure the axes, and we would have tested a hundred axes (not six). But even still, we walked away with a few good findings:

The Results

INTEGRITY (IN WOOD)

Of our three wood samples, one had good (straight) grain structure, and it far outperformed the other wood samples, proving that the grain orientation in wood axes matters. The plastic axes (predictably) performed consistently (i.e., they all have the exact same internal structure).

COMFORT

In addition to static loading, we tested the dynamic response of the axes. Each axe was cyclically loaded to a consistent deflection before being unloaded. The dynamic response created a hysteresis loop, whereby the load upon the return was less than the load up to the point of maximum deflection (that energy was stored in the axe on the return). I was pleased to see that the wood handles proved to have a larger loop, which indicated they are more likely to have a damper feel than the plastic axes, proving my personal theory that plastic axes send much more reverb through the helve when you make contact (i.e., wood axes are more comfortable).

STRENGTH

Although our best wood axe failed at a lower load than any of the plastic axes, it was nonetheless stronger and stiffer (before that moment of failure) than any of the plastic axes. If that's confusing, just imagine the difference between bending a toothpick vs. a plastic straw.

Bottom line:

There are many more tests that could be conducted, and until then I remain convinced that wood is, by far, superior to plastic.

The axes were tested as static cantilevered beams (above) with the load applied to the axe-head, while the end of the helve was constrained to best represent the dynamic loading of the head striking a log. We subjected each axe to a maximum static force of 600 to 850 Newton (150–180 pounds), well beyond the force that you or I could apply.

7. Dynamics of the Strike

The intended use of an axe is seemingly simple: You swing it toward a target, and upon impact, the target is fractured. Felling, limbing, bucking, and splitting all amount to fracturing, though each requires a particular swing in order to achieve the ideal impact. When you're learning to swing, it's easy to presume that your swing requires maximum output of strength and force. With experience, this presumption will reveal itself as flawed, and your swing will feel much more natural, and lo and behold, as you calm down, you start hitting the target with greater frequency and efficiency. At that point it clicks: The axe can and should do the majority of the work for you.

Your axe is essentially a speed-multiplier lever. The common idea of a lever, such as the hypothetical lever Archimedes would use to lift the world, is that a fulcrum is placed between the load (the thing being lifted or moved) and the effort (something pushing down on the other

The speed-multiplier lever

end). The lever makes the job easier, and we're able to lift heavier things than our bodies would typically allow. A speed-multiplier lever is different. The effort is applied between the fulcrum and the load, meaning you're actually using more energy by swinging an axe than you would by simply swinging an axe-head at arm's length. This emphasizes the importance of fitting an appropriately weighted axe bit to the user's strength and stamina. Nonetheless, the faster the axe-head's impact, the greater the force, and the better the end result.

Getting the swing of it

But keep in mind there's always a fine line between fast and too fast, and you can let gravity do much of the work. With your base hand planted at the butt of the helve, and your forward hand up by the helve shoulder, you can lift the axe up over your head, only needing to overcome the weight of the head, and by swinging the axe forward while sliding your forward hand down to meet the base hand, guide the axe forcefully into its target. Any energy you input into the helve during its descent accelerates the head well past the terminal velocity achieved by gravity (see page 145).

On impact, the kinetic energy of the axe-head in motion is transferred into the wedging action of the bit. Let me take a moment to reiterate the importance of a sharp edge at this point. A sharp edge will bite directly, transferring the force of the swing into the intended cut (if the chopper's aim is true). A dull edge does not bite as well,

Stay sharp

potentially allowing the grain of the wood to deflect the force of the axe in unintended directions. Worse yet, the dull edge may not bite at all, and the force of the swinging axe will follow through into dirt, your boot, or empty space. Additional energy will be spent in arresting the wayward swing of the axe, making your day a lot longer and a lot less fun. The intent of the chop is either to split along the grain of wood, in the instance of splitting firewood, or to sever the wood grain, in the case of felling or limbing. In both cases, control and accuracy are your friends. Directing the axe edge repeatedly to the same spot or along the same line will allow the axe to work for you and allow you to build necessary muscle memory. Remember, the axe is a combination of simple machines designed to offer mechanical and energetic advantage. With consistent aim and a sharp edge, the design of the axe achieves its maximum efficiency. While cutting across the wood grain, a dull edge will crush rather than slice, resulting in a loss of impact energy and throwing smaller chips.

When splitting firewood, you're at risk of allowing the repetitive, contemplative nature of the task lull you into routine, which requires you to design safety into your technique. The proper wood chopping arena is a patch of level ground with a large, durable log round at the center. The surface of the log round should be close to parallel with the ground, to avoid time spent fussing to balance a target log. The ground should be clear of obstruction, and occasionally cleared of all chopped wood to prevent any stumbles with an axe in hand. Consider the full swing of the axe and make sure there are no low branches, gutters, or shed roofs that might intersect with the arc of the swing. If any obstructions come close, adjust either the obstruction or your location, not your swing. Once the swinging begins, the fervor of chopping takes over and you'll quickly forget potential obstructions.

Safety and obstructions

If you poorly judge the distance to the target, there's a risk of striking either too far to the toe of the bit or too far to the heel. Striking too far to the heel is a common mistake, often the result of a timid swing, which typically results in the shoulder of the helve striking the target. This is called overstrike, and is not just the mark of an unskilled

Overstrike

chopper—it's a sure way to end up learning how to re-helve an axe, since it can split your helve. Some axe bits are constructed with a steel collar that covers the area of the helve most susceptible to overstrike. Collars can make the helve hanging process more tedious, however, and don't excuse poor form. If overstrike occurs, be sure to examine the entire helve for any long, longitudinal cracks that may foretell a catastrophic failure. The helve shoulder is durable and can bear a few mistakes, but can't withstand overstrike forever.

Striking too far to the toe is even worse, as the axe is suddenly cleared of any obstructions until it finds the ground or a boot. Find the ground, and you might face a few hours of filing and resharpening to take the chip out of your prized bit. Find the boot, and your day just got a whole lot different.

When you chop into a piece of wood, it's impossible to anticipate every single outcome every single time. That's why you need to plan ahead, and map your next steps. If in **Plan ahead** doubt, I will literally take slow practice swings on a chopping block, just to be 100 percent sure I know where my axe is going and what might be in its way. You would be surprised what damage you can do if, in full swing, your axe gets hung up on a seemingly harmless little branch or vine.

Think of the axe like a well-trained dog. An obedient dog is responsive to your inputs, or "commands," while an unruly hound spells danger. Obedience requires patience and training, just as the axe requires care, maintenance, **The** and practice. With practice you will build muscle memory, **obedient** and chopping wood will feel as natural as riding a bike **axe** or skipping rope. The more time spent swinging the axe, methodically practicing technique, and keeping it clean and sharp, the less likely it is to come back and bite you.

THE SWING

Master accuracy before speed and force. You don't need to swing as hard as you think. And always remember, a sharp axe is the most accurate axe.

Keep the path of your axe well clear of overhanging branches and obstructions.

Your top hand is loose and will glide the length of the shaft. Your bottom hand is the anchor of your swing.

Don't lock your wrists. The snap of the wrists accelerates the moving axe and gives it that necessary whip.

From the minute you raise your axe, keep your eyes fixed on the target. As you lower the axe, imagine cutting through the target, not just into it.

Keep your shoulders square and level.

BUILD MUSCLE MEMORY
The key to becoming an effective chopper is practice and repetition. Swinging the oldest tool known to humankind is in your DNA. Tap into that ancient muscle memory!

PLAN AHEAD
Do a mental walk-through of your swing and its intended path. Trace the actual path of the axe if you need to. Every swing should land where you intend it, but know that the chances of that happening are not guaranteed.

Keep your knees slightly bent.

Keep a wide and stable stance. Know what's underfoot.

Wear the right boots.

Keep your swing straight, and the axe in this channel on the downward stroke.

So here I am in outfit green
Crossed axes and saw
Felling trees from morn till night
To help win this blinkin' war
I've been at it now a year or more
It's just the life for me
Until the bells of peace ring out
Up the WTC.

From "Ode to the WTC"
published in *Meet the Members: A
Record of the Timber Corps of the
Women's Land Army*, 1944

Ode to the WTC

Officially, they were known as recruits to the Women's Timber Corps, a civilian agency established by the British government to supply the timber industry with new workers after its numbers were depleted by conscription. (Timber was a crucial resource for a nation at war. It provided materials for housing and aircraft, and it provided important components for coal production, such as pit props.) But they were commonly referred to, and still are, as "lumberjills."

I think it's an oddly gendered term considering their duties were the same work done by the men they replaced: felling and chopping wood, then hauling it to wherever it needs to go, sometimes by putting it on your shoulder and taking it there.

The surviving photographs of the lumberjills show smiling women hauling and chopping, bucking and sawing. But these were mostly teenagers, many of whom were out in the country for the first time. For most of them, joining the corps amounted to a kind of evacuation from the cities and towns they grew up in, now enemy targets, to an alien landscape of fields and forests, which came with a different kind of danger. I'm sure many of their smiles were genuine, but I'd guess just as many of those smiles faded as soon as the photographer put down his camera.

They performed every task a male forester performed. They measured, felled, limbed, bucked, and hauled. And they endured every indignity a male forester endured: cold, rain, and wind. They lost fingers and limbs (both out in the field and in the sawmill). And they were killed by falling trees.

From 1943 to 1946, as many as 8,500 women participated in the program. Surviving members speak with great pride about how they learned to fell 100-foot-tall pines and how they marveled at the sight of the topmost branches falling away from the forest canopy and the trunk thumping against the earth. In her remarkable book *Lumberjills: Britain's Forgotten Army,* Joanna Foat features extensive reminiscences of former Timber Corps recruits, including this one from Peggy Conway about her first day on the job:

> So there I was at 8 a.m. . . . and an open-backed wagon arrived and tipped out a load of wild-looking men, who climbed over the fence with the axes and crosscut saws, swearing vociferously. A bit of a shock to a nicely brought up Methodist girl!

That passage indicates that this was never "all-female" work, despite what the photos would suggest. They sometimes worked alongside men—as partners and, to a great extent, equals.

8. Dynamics of Wood

One of my favorite proverbs is "The tree remembers what the axe forgets." It's so easy . . . I couldn't have put it any better myself; words of wisdom to hold near and dear. It's so easy for there to be a disconnect between the log that's burning bright and beautifully in your fireplace and the tree that gave its all. This chapter is designed to be a simple but necessary primer, and by no means a thorough discourse, that will help draw that connection, improve your ability to work with wood, and make you better with an axe.

Every plant to have evolved after the mosses and liverworts—ferns, conifers, flowering plants, etc.—has a vascular system. It's a core feature that allows trees to reach the impressive height of the Giant Sequoias: over 300 feet (91 m). Plants similarly have two types of tubes like we do, but unlike veins and arteries, which both pump blood but in different directions, the vascular tubes of plants have separate functions.

The vascular system

The two types are called xylem and phloem. Xylem transports water from the roots (where it's absorbed) to the leaves (where it's used for photosynthesis), in a unidirectional fashion. Phloem moves the sugar product of photosynthesis from the leaves to the area of the plant where it is needed, in any direction. The wood of a tree is a bundle of xylem, stacked together in a cylindrical stem, while the phloem forms in the bark, underneath the tough exterior layer. It's easy to understand why "girdling" (removing the bark) quickly kills a tree, by disrupting its vasculature.

Referring to the same cross section, you can identify several distinct areas. The dark central area is the heartwood, sometimes with a pithy core. This area is "dead"—its vasculature is no longer active for transport of water or nutrients, but it continues to provide structural support. While heartwood is naturally durable and decay resistant, the very core, or pith, of some trees remains spongy and soft. The lighter sapwood is active, "living" vasculature, with working xylem.

Heartwood, sapwood, cambium

Between the bark and the sapwood is a thin layer called the cambium, where growth happens. This is how tree growth extends outward, resulting in the formation of annual rings. The bark is the exterior, containing the phloem while providing a durable layer of protection from disease and invasion. Each individual growth ring has a pattern of light and dark as well, distinguished by the rapid period of growth in the spring and summer, followed by slow growth and dormancy in the fall and winter.

Wood is commonly distinguished as hardwood or softwood. But these designations are misleading. The definition of these categories lies not in the actual hardness of the wood but in the trees' places in evolution. The hardest softwood is actually harder than the softest hardwood. Hardwoods are mostly deciduous. And softwoods are mostly coniferous; they are typically used for economical applications, especially construction, due to the rapid growth of softwood trees and the ease of milling.

Hardwood and softwood

The structure of hardwood and softwood is inherently different, resulting in distinct behavior when chopped or burned (see page 92). Softwood, being the more ancient of the two, has a simpler construction with long, thin, tubular cells. The hardwoods are much more complex, with less-fibrous, less-uniform cell arrangements, and a denser construction. Softwoods are typically lighter and less dense than hardwoods. They split very easily given their simple construction, and handle easily with their lighter weight. Hardwoods are much tougher to split, given the disruptions in their cellular structure caused by vessel elements, and a much more tightly packed cellular arrangement. The effort is well worth the payoff, though, because hardwoods are vastly superior for burning.

Softwoods also contain large resin canals that hardwoods lack. While pine or cedar is fairly easy to chop and ignite, they don't make a good fire. The lower density and high resin content of the wood means it burns hotter and faster, with lots of crackles and pops. The burnt resin will lead to a buildup of creosote in a chimney, which can have potentially dangerous consequences, or at the very least lead to the need for more frequent chimney cleanings.

Due to their ease of ignition, softwoods can be chopped into kindling and kept by the stove or fireplace to help get a fire going quickly. As long as softwoods are not the bulk of the wood fuel, they're excellent for starting or reanimating a fire. Hardwoods, the denser of the two, take longer to get up to the temperature necessary for ignition, but once lit will burn longer with steadier heat. Whether burning wood for heat or cooking, the hardwoods will pay off considerably when it comes to heat produced per cubic foot.

Heat produced is typically measured in British Thermal Units (BTUs), which is the heat required to raise 1 pound (455 g) of water by 1°F (0.56°C). For a quick comparison, applewood produces about 26 million BTUs per cord, while white pine produces about 14 million BTUs (see page 92).

BTUs

Green wood, or wood that's been freshly chopped, is the most difficult to burn, and will give off loads of steam and smoke if added to a fire. This is why firewood is seasoned, or dried, over the course of at least six months.

Seasoning wood

If only green wood is available, the wood can be "feathered" by making a series of slashes along the length of the wood. This maximizes surface area, to speed the process of evaporation. As a general rule, the denser the wood, the longer it takes to dry, so cedar kindling will be ready much faster than ash firewood. Drying occurs fastest at the ends of the logs, which is why firewood is stacked to allow airflow across the cut ends. The majority of the water bulk will evaporate quickly, but the water trapped in the wood cells takes time to escape. The firewood pile should be covered as protection from rain and snow to prevent the logs from reabsorbing moisture. Drying time will vary depending on climate, but even in northern regions, wood cut in spring should be dry and ready to burn that winter. If a moisture probe is used, firewood is ready to burn when the moisture content drops to 20 percent.

The chemistry of fire

Wood is comprised of primarily cellulose and lignin. Cellulose is basically the molecular fiber of wood, while lignin is the natural "glue." Also present are various other chemicals, such as oils or tannins, that give each wood species its distinctive color and aroma. When wood burns, it's not the wood matter that's going up in flames, but the gases released by the wood. This is why we observe the shape of a campfire with flames licking upward, rather than wood brightly radiating in the shape of the log. Wood begins burning when it reaches the temperature of ignition, around 572°F (300°C). The cellulose begins to decompose under heat, or pyrolyze (pyro = fire, lyse = cut or break). As the chemical bonds of cellulose break, flammable gases are expelled from the wood. These gases spontaneously combust in the presence of heat and oxygen, releasing more heat, causing more wood to burn and more gases to combust.

The process is a feedback loop, which is why fires persist as long as there's fuel—campfires don't suddenly cut out unless deprived of fuel, air, or heat. As the wood pyrolyzes, it's reduced to char, which is essentially pure carbon, and how we get charcoal. Minerals present in the wood, such as calcium and potassium carbonate, that do not burn materialize as ash. Molecules that are released as gases but don't combust form a cloud of particulates we call smoke. Charcoal, which is mostly lacking in secondary compounds, does not smoke for this reason. The chemical compounds in smoke are often toxic, which is why woodstoves that encourage complete combustion or open-air campfires are preferred to fireplaces.

Choosing the right wood to burn is as important as being able to start a fire at all. Most folks know it's best to reach for hardwood over softwood, which is a good general rule, but exceptions exist. For the most part, hardwoods (from deciduous, broad-leaved trees) are denser woods without the resin content of softwoods (from evergreen conifers). As a material, wood generally burns at the same heat pound for pound, but not all woods have the same burn quality. The denser hardwoods are harder to light, can be harder to split, and, of course, weigh more, but burn with more heat per log. This makes them more ideal for the bulk of the fuel, as wood is usually measured by the cord (volume), not weight. Get a fire going, and hardwoods such as oak and ash will sustain you through the winter.

Softwoods, on the other hand, are much less dense, thus require more logs to produce the same amount of heat. They burn quickly and tend to be smokier and spark more often, leading to more air pollution in your home (if burning in a fireplace) and more buildup in your chimney. If

Hardwoods as fuel

THE FIREWOOD CHART

SPECIES	WEIGHT (LBS./KG PER CORD) DRY	HEAT PER CORD (MILLION BTUS)	EASE OF SPLITTING	SMOKE
ALDER	2540 / 1152	17.5	Easy	
APPLE	3888 / 1764	27	Medium	Low
ASH, GREEN	2880 / 1306	20	Easy	Low
ASH, WHITE	3472 / 1575	24.2	Medium	Low
ASPEN, QUAKING	2160 / 980	18.2	Easy	
BASSWOOD (LINDEN)	1984 / 900	13.8	Easy	Medium
BEECH	3760 / 1706	27.5	Difficult	
BIRCH	2992 / 1357	20.8	Medium	Medium
BOX ELDER	2632 / 1194	18.3	Difficult	Medium
CHERRY	2928 / 1328	20.4	Easy	Low
COTTONWOOD	2272 / 1031	15.8	Easy	Medium
DOGWOOD	4230 / 1919	High	Difficult	
ELM, AMERICAN	2872 / 1303	20	Difficult	Medium
FIR, DOUGLAS	2970 / 1347	20.7	Easy	High
FIR, WHITE	2104 / 954	14.6	Easy	Medium
HACKBERRY	3048 / 1383	21.2	Easy	Low
HEMLOCK	2700 / 1225	19.3	Easy	
HONEY LOCUST	3832 / 1738	26.7	Easy	Low
LARCH (TAMARACK)	3330 / 1510	21.8	Easy-medium	
MAPLE, OTHER	3680 / 1669	25.5	Easy	Low
MAPLE, SILVER	2752 / 1248	19	Medium	Low
MULBERRY	3712 / 1684	25.8	Easy	Medium
OAK, BUR	3768 / 1709	26.2	Easy	Low
OAK, RED	3528 / 1600	24.6	Medium	Low
OAK, WHITE	4200 / 1905	29.1	Medium	Low
PINE, PONDEROSA	2336 / 1060	16.2	Easy	Medium
PINE, WHITE	2250 / 1021	15.9	Easy	
PINYON	3000 / 1361	27.1	Easy	
POPLAR	2080 / 943	Low	Easy	
RED CEDAR, EASTERN	2060 / 934	13	Easy	Low
RED CEDAR, WESTERN	2632 / 1194	18.2	Medium	Medium
SPRUCE	2240 / 1016	15.5	Easy	Medium
SYCAMORE	2808 / 1274	19.5	Difficult	Medium
WALNUT, BLACK	3192 / 1448	22.2	Easy	Low

Source: Utah State University, excerpted from "Heating With Wood: Species Characteristics and Volumes," by Michael Kuhns and Tom Schmidt.

SPARKS	COALS	FRAGRANCE	OVERALL QUALITY
Moderate	Good	Slight	
Few	Good	Excellent	Excellent
Few	Good	Slight	Excellent
Few	Good	Slight	Excellent
Few	Good	Slight	
Few	Poor	Good	Fair
Few	Excellent	Good	
Few	Good	Slight	Fair
Few	Poor	Slight	Fair
Few	Excellent	Excellent	Good
Few	Good	Slight	Fair
Few	Fair		
Few	Excellent	Good	Fair
Few	Fair	Slight	Good
Few	Poor	Slight	Fair
Few	Good	Slight	Good
Many	Poor	Good	
Few	Excellent	Slight	Excellent
Many	Fair	Slight	Fair
Few	Excellent	Good	Excellent
Few	Excellent	Good	Fair
Many	Excellent	Good	Excellent
Few	Excellent	Good	Excellent
Few	Excellent	Good	Excellent
Few	Excellent	Good	Excellent
Many	Fair	Good	Fair
Moderate	Poor	Good	
Many			
Many	Fair	Bitter	
Many	Poor	Slight	Fair
Many	Poor	Excellent	Fair
Many	Poor	Slight	Fair
Few	Good	Slight	Good
Few	Good	Good	Excellent

softwoods such as cedar or pine are harvested, they make for great fire starters. Split up into kindling and keep a bundle by the stove for a lively, crackling start to every fire, or to renew the coals from the night before. However, there are some deciduous hardwoods such as cottonwood or quaking aspen that don't have the density or clean-burning properties shared by most hardwoods. Conversely, Douglas fir is an abundant softwood in the American West that burns quite well. As species vary across the world, there will be optimal choices for every county. When I bought my property in the Catskill Mountains, one of the first things I did was hire a local forester to walk the property with me, and this small investment was one of my best. He taught me almost everything I needed to know about the trees on my property, including the best practices for not just harvesting them, but for being a steward of the land.

Softwoods as fuel

Keep an eye on the chimney while the fire roars. Until it's really going, almost any fire will smoke, but once it's caught, a good hot fire burns cleanly, without heavy smoke billowing out the chimney. The hardwoods in the "Best" and "Good" categories will be less prone to smoky fires, but any large logs left smoldering in the stove will smoke profusely. Get the firebox up to temperature and the fire will be easier to maintain and will burn cleaner. Be sure to burn dry wood only. With excessive moisture in the logs, heat will be absorbed in the process of evaporating the water trapped inside, rather than being converted to combustion heat. The resulting fire will be smoky as well. Make sure the wood has had sufficient time to season. When purchasing firewood, handle the wood to be sure it isn't excessively heavy, and look for cracks at the ends of logs that form as wood dries. In time, you'll find what woods are abundant in your area, and what splits and burns to your liking.

Best practices for burning

FIREWOOD FACTS

MOISTURE CONTENT

Firewood should be dried until its moisture content is less than 20 percent. To know that you'll need a wood moisture meter (available for less than $50). Burning wood with a moisture content higher than 20 percent is not only difficult, but it won't burn cleanly and can result in dangerous creosote buildup in your chimney.

SEASONING

It may take six to twelve months (possibly more depending on your climate) to season your wood, so plan in advance. I split my wood when it's green: it's easier to split and dries faster.

PROTECTING

Keep your wood pile covered, but not completely. It's important to have airflow, and somewhere for the moisture to escape. Try and keep your wood piles in sunny spots with good airflow.

STACKING

There are more elaborate methods (i.e., a Holz Hausen stack) but I like to keep my stacks simple. I stack my logs neatly into woodsheds. Using the logs, I build columns to support the stacks by layering the end logs horizontally, and the next layer perpendicular (and so on). This method known as "cribbing" creates two columns that will act as bookends to the stack, and allows you to stack a higher and sturdier amount of logs between the two columns.

BURNING

Good seasoned wood should burn cleanly (i.e., without much smoke). A wood stove gives off at least three times as much heat as a fireplace. No matter what you use, get your chimney cleaned at least once a year.

THE CORD

Firewood is sold in units called "cords." The depth of a face cord (also known as a "rick") can vary and is a more informal measure than a full cord.

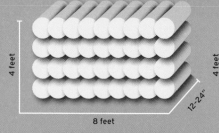

4 feet
8 feet
12-24"

FACE CORD (cubic feet varies)

4 feet
8 feet
4 feet

FULL CORD (128 cubic feet)

Buying

Inside the storeroom at the
Wetterlings Forge in Storvik, Sweden.

9. Buying New

In a perfect world, you would spend an entire day at your
favorite axe maker, watching the head of your axe being
forged, tempered, and sharpened. You'd see them shape
your helve and hang it on your head. Once it was complete,
you'd hand the maker a few crisp twenties, and off you'd go.
The reality is, you will trudge down to your local hardware
store, or more likely turn on your computer and buy your
new axe online.

No matter what your source, there are a few things to
look out for, some that should automatically disqualify a
new axe from purchase: any axe with a helve made of anything other than wood, any axe to which epoxy has been
applied to the eye of the axe.

Fiberglass and plastic composite helves are unneces-
sarily difficult and messy, if not impossible to replace, and
don't provide the same kind of comfort when you strike as
hickory. Shorter axes and hatchets may have a full steel
helve integrated with the head. This means the impact of
the strike will ring through the head and down the helve
like a tuning fork, transferring the vibration directly into
your bones.

Fiberglass
and plastic
helves

As has been the trend for most tools over the years,
the axe has been given the appearance of requiring
less maintenance, with the result being a tool more difficult
to maintain overall. This is the case with helves completely

finished with varnish or filled with epoxy at the eye. Varnish may seem to make sense, but it increases the friction along the helve, guaranteeing an unpleasant swing. An epoxied eye may totally seal the end grain and fuse head to helve, but it's a nightmare to remove. While these two treatments completely weatherproof a stick of hickory, they also reduce the usability and performance of the tool, and make a perfectly biodegradable material perfectly toxic.

Then what does the ideal new axe look like, and where is it found? The helve is American hickory lightly finished with linseed oil. The helve grain will be parallel or close to parallel with the head. Grain direction is less important for shorter axes, but it's worth a check. The helve stock should have been graded in the factory, but sometimes a dud slips through—look for any knots or defects while you're at it.

The steel should be medium- to high-carbon steel, drop forged in an open or close die. Again, the higher carbon content is what you need for a performance tool. It will be a tool that requires attention and respect—leave it wet after a day of splitting and it will rust. Chop at roots and it will chip. Steel that is forged in the open-die process results in a more handcrafted, singular end product. The closed-die process lends itself to consistency, and a more uniform end product (see page 52). Depending on your needs, budget, and personal preference, there is a case to be made for each process, and I wouldn't argue one makes for a better steel than the other: There are so many other factors at play.

In the end you have the option of buying either a new axe or an old one, and I want you to think long and hard about that decision. Shiny and new can be so much more tempting than rusty and old. In the next chapter I'll attempt to reverse that mindset, but before that, let me just say why and when it pays to buy a new axe.

In most cases, an old axe will require much more of you. For starters, you will probably have to hang a new helve on your axe, and you may even have to completely restore it, which means time . . . and vinegar baths and bastard files, etc. There are obvious advantages and disadvantages to all that legwork, and out of the box a new axe is good to go with a little shaping, while an old axe is a project. Most sellers of old axes know nothing about them, and if they do, they're going to charge you for it. When you buy new, you have the chance to ask important questions like: Where was the axe made, what type of steel is it, where is the steel from, what's your guarantee? These are all questions that a seller of an old axe will likely not be able to answer. Likewise, there are ongoing advantages of buying new, such as customer service, and the ability to speak to someone if you have questions. There are an increasing number of terrific options for buying a new axe, and many of my favorite makers are small and need our support.

A special note about buying axes (from a guy who's sold thousands of axes): If you're buying online, don't be afraid to place your order over the phone, and ask if they can personally go into their inventory and pick out your axe. Tell them (respectfully) exactly what you want: the straightest possible grain, your preference for a light helve (sapwood) or a dark helve (heartwood). We had customers who made these requests, and they got what they wanted. If, when you receive your axe, you don't like what you see, then send it back, hopefully at the seller's expense. If you're in a store, ask to see all their axes, the entire in-store inventory, and then pick the best of the litter. Not all axes come with guarantees, and if they don't, then good customer service is your best ally. If something goes wrong with your axe, contact the maker before you draw any conclusions, show them pictures, and walk them through the issue at hand.

What to ask when you buy new

The Poet's Axe

This is an excerpt from one of my favorite poems, "The Axe-Helve." In it, Robert Frost, a Vermonter who knew how to swing an axe, assumes the voice of a narrator splitting wood on his own property. He's chopping away until one of his swings is arrested by his neighbor, a man named Baptiste who he doesn't know all that well. The narrator, naturally, wonders if this is an aggressive act. Turns out the neighbor only wants to say that the axe's helve is of inferior quality—"made on machine"—and could snap. Baptiste invites the narrator to his house, where he'll give him a much better axe. The danger of grabbing at an axe mid-swing aside, this poem highlights a number of important points about the axe: (1) If the axe doesn't have integrity, don't use it. (2) We have a great responsibility to our neighbors, even when (and perhaps especially when) they don't think they need our help. (3) There should be more axe poetry.

I've known ere now an interfering branch
Of alder catch my lifted axe behind me.
But that was in the woods, to hold my hand
From striking at another alder's roots,
And that was, as I say, an alder branch.
This was a man, Baptiste, who stole one day
Behind me on the snow in my own yard
Where I was working at the chopping-block,
And cutting nothing not cut down already.
He caught my axe expertly on the rise,
When all my strength put forth was in his favor,
Held it a moment where it was, to calm me,
Then took it from me—and I let him take it.
I didn't know him well enough to know
What it was all about. There might be something
He had in mind to say to a bad neighbor
He might prefer to say to him disarmed.
But all he had to tell me in French-English
Was what he thought of—not me, but my axe;
Me only as I took my axe to heart.
It was the bad axe-helve some one had sold me—
'Made on machine,' he said, ploughing one grain
With a thick thumbnail to show how it ran
Across the handle's long-drawn serpentine,
Like the two strokes across a dollar sign.

An excerpt from "The Axe-Helve" by Robert Frost

10. Buying Old

Best Made worked closely with Council Tools, a fourth-generation forge in North Carolina, to produce one of the first premium "modern-era" American felling axes. At the time, our axe was the most expensive you could buy, on par with the incredible specimens that had already been coming out of Sweden. In the early days, some of our customers would reach out to tell us how much they wanted a Best Made axe, but quietly asserted they couldn't quite justify the expense. They'd reminisce about the axes of their childhoods, invariably the ones an ancestor had swung. Little did they know there was a lot of life left in those axes; there was no reason why they couldn't be brought back into service and be as good (or better) than any new axe on the market. And so that is how Best Made got into the axe restoration business.

A demand for restoration

We were obviously not the first to restore an axe, but to my knowledge Best Made was the first retail brand to offer axe-restoration classes. Early pioneers of "restoration-education," we hosted our first class in our 180-square-foot (55-square-m) workshop in Lower Manhattan, and were soon on the road, heading out across the United States and to parts of Europe to teach folks how to breathe life into derelict but undeniably vital and beloved axes.

An axe maker once told me that a million (new) axes are bought and sold every year in the United States. Multiply that by a hundred-plus years and you have a serious accumulation of axes. Most of these axes are not up for grabs—they're likely hanging in toolsheds in varying states of use or neglect. The quantity of old, used axes far outweighs brand-new or unused axes. But they can be hard to find.

The art of
axe hunting

Since I started Best Made, demand for old axes has soared. In ten years, the prices have skyrocketed, and the selection has thinned. Finding a good old axe is entirely possible, but it takes a little more legwork and patience than it used to. Axe hunting in-person (that is: not online) is best executed with a car. Although I have flown around the world with axes in my checked luggage, I still prefer the restrictions of my Subaru to the TSA. Junk shops, flea markets, antique stores, and dealers that specialize in antique tools are ideal hunting grounds. Nothing beats the dusty old shelves of Liberty Tool in Maine, but sadly, I can only get that way about once or twice a year. I'm not ashamed to admit that I've sourced a big part of my collection on eBay. The selection is almost limitless and geographically far-reaching, the turnover is frequent. And I can take my time and make a decision from the comfort of my favorite armchair.

Hanging
an axe

Whether it's a dusty shelf in the backroom of a roadside flea market, or a pixelated product photo online, when you find the old axe of your dreams, there are always a few realities to embrace. Either the axe will have a helve or it will not, and if it does it is likely to be dried up, cracked, or poorly hung. And despite their valiant service once upon a time, most old helves are sadly the first thing to go.

You have to ask yourself: Is this an axe you want to use or just collect and admire? If you want to use it, then you'll want to cut off the helve and hang a new one, and in doing so you are making your own axe (and that is reason enough to buy old). If this axe is the next addition to your personal museum, then I advise leaving the helve intact. No matter how worn or damaged it might be, it should have some good stories to tell.

Zooming in I will look for vestiges of a maker's mark (see page 110) or some design element that could provide some background and provenance. Most old axes out there, especially the lesser priced, won't have this type of indicator, and if they do they'll be hidden under the rust and grime. I recently found a dream axe on eBay; it was the elusive and highly collectible Plumb Autograph. This axe had beautiful, clean, illustrative markings, and it was priced to sell (suspiciously so). And so, after scrutinizing the photographs, I discovered it had an alarming crack down its cheek. Was this going to be the greatest axe-restoration project of all time? I reached out to a blacksmith in Upstate New York and asked if there was any hope of restoring this beauty. Sadly, he said it was a goner; a weld would have completely compromised the temper of the steel. We'd be better off reforging the entire head, and that would have meant losing all those beautiful markings. I bought the axe anyway, and now it's an overachieving paperweight.

Reading the maker's mark

Sellers often take the restoration upon themselves, and this usually doesn't end well. When buying an old axe, you should expect it to be dirty, covered in pitch, dull, and almost certainly rusty. It's so easy to overlook a rusty tool and assume it's done for, but this couldn't be further from the truth. Rust, even a lot of rust, can easily be dealt with (see page 185). More and more sellers know this, but they rush the restoration process using heavy abrasives and invasive grinders that may possibly compromise the integrity of the steel. What drives me crazy is when they polish the axe to a mirror finish thinking they will command double or

Rust is not the end of an axe

triple the price. Thank you very much, but I will be the one who decides if I can see my reflection in my axe.

The price factor

Price is predictably a determining factor in buying an old axe. I give myself serious price constraints because it forces me to be resourceful and it adds some thrill to the chase (throwing money at an axe is cheating). At the time of the publication of this book, I can source a great axe for under $20. This may not get me a collector's elusive gem, but it will almost guarantee an axe that I can put to good use, and in time it will become my own personal gem. And look, if in the end it doesn't work out, if the axe isn't everything you had hoped for, then you've got yourself a great new paperweight, too.

Look at the lines

When evaluating an old axe, look closely at the lines: Are they deformed, compromised, overworked, abused? Or are they tight, clean, and crisp? Watch out for axes with polls that are mushroomed out—they've likely been used as sledgehammers. Axes are meant for cutting wood, not pounding steel, and the polls of most axes were not tempered like the bits; they're softer and less resilient. Damage to the poll can just be cosmetic, or it can cause serious damage to the eye of the axe, and that's a deformity that no restoration can save.

Read the bit

My father used to sharpen his little Swiss Army pocket-knife on our farm's high-speed electric grinder, and within no time the blade would be honed down to a toothpick. Axes, like knives, should be sharpened by hand. This fundamental truth is something that my dear old father, and many others, have yet to grasp. Electric grinders remove too much material too quickly, and they can heat the blade past the point of no return, ruining the temper of the steel. An axe that has been run through the grinder usually has an overly rounded bit, whereby the toe and the heel have basically been ground right off. Just as you want a square, clean poll, you should be looking for tight clean lines in your bit—toe, heel, and in between. I should note that electric grinders are used effectively by professionals who use the right machines, abrasives, jigs, and sharpening techniques. But for the rest of us: We can do just as well with a bastard file and a sharpening stone.

Kintsugi is the ancient Japanese art of restoration, specifically the use of a gold or silver lacquer to mend broken pottery, thereby transforming worthless shards of porcelain into a functional, beautiful vessel. I think of axe restoration as a similar pursuit, just as timeless and noble as Kintsugi. The thrill of reaching into a brown and briny broth and pulling out an axe that has been bathing for twelve hours in vinegar, and then seeing the effect of acid on rust, will never get old. There is an almost miraculous reward in the outcome. Through restoration, we can't help but develop a deeper connection to the axe: We make it better and we make it our own. When I swing an axe that I've restored I feel much closer to the ancient continuum of the oldest tool known to humankind. I may not know the exact history of that axe, but I can imagine: who has swung it, where it's lived, and what else it has seen.

The thrill of restoration

The lifespan of carbon steel is a moving target, but if cared for, your axe should last as long as almost anything you can possibly possess, and that means many many (many) lifetimes. Who will be swinging your axe in five hundred years? The axe will always be there, it's just you who is passing through. It's your duty to maintain and, if need be, restore these tools while you have them, to use them and keep them functioning, and to ultimately leave them better than you found them.

The legacy

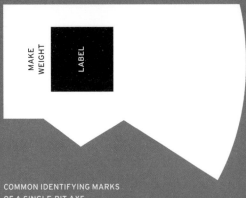

MAKE WEIGHT LABEL

COMMON IDENTIFYING MARKS
OF A SINGLE-BIT AXE
(a number under the poll may
indicate the year)

11. Identifying Marks and Labels

On your hunt for a rare old axe, I hope you're lucky enough to stumble across a specimen with a mark (be it a name, an etching, or, best of all, a label). At the dizzying height of the early-twentieth-century axe market, these markings were often an axe's best chance at standing out. As you can see in the following pages, evocative names were given to axes, and they were emblazoned with fancy lettering and illustrations that beckoned the Paul Bunyan deep within the axe seeker. I discovered this glorious graphic history in the early months of Best Made, and the flair of these old makers was then, as it is now, a constant inspiration. There were once some maker's standards for how these marks were applied (see above), although tragically many axes were made with no markings at all. The following are a selection of my favorite labels from the collection of Nick Zdon.

HAND MADE
CHARCOAL TEMPER
MANUFACTURED BY
NORTH WAYNE TOOL COMPANY,
HALLOWELL, MAINE.

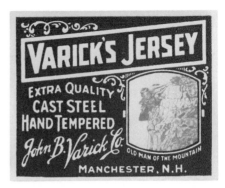

VARICK'S JERSEY
EXTRA QUALITY
CAST STEEL
HAND TEMPERED
John B. Varick Co.
OLD MAN OF THE MOUNTAIN
MANCHESTER, N.H.

Old Time Axe
Forged And
of the Carefully
Best Tempered
Steel by Hand
by James H. Mann near
Lewistown, Pa., U.S.A

MANUFACTURED FROM
SILVER
BEST THE STEEL
WETMORE
AXE
EMERSON & STEVENS MFG. CO.
OAKLAND MAINE

WASHINGTON
TRADE MARK
FORGED FROM HIGH GRADE TOOL STEEL
REGISTERED
TOOLS
WARRANTED TO HAVE NO SUPERIOR.

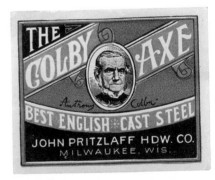

THE
COLBY AXE
Anthony Colby
BEST ENGLISH CAST STEEL
JOHN PRITZLAFF HDW. CO.
MILWAUKEE, WIS.

The
Niagara
REGISTERED
BEST HAMMERED
CAST STEEL
MANHATTAN AXE CO.
NEW YORK

EXTRA
HAND HAMMERED
VERY BEST AMERICAN
BIT STEEL
EVERY AXE REFINED
AND TEMPERED BY HAND
NORTH WAYNE TOOL CO.
OAKLAND, MAINE

CUTTER, WOOD & SANDERSON CO.
CUTTERWOOD
HAND FORGED AXE
CAMBRIDGE, MASS.

THE
PIONEER
AXE
LEWISTOWN, PA. U.S.A.

OLD YANK
HAMMERED
AXE
ALL HAMMERED
NOT
DROP FORGED
NEW ENGLAND HANDLES
THOMPSON, CONN.

SPECIAL GRADE
SOLID STEEL
Wm ENDERS
OAK
LEAF
WALDEN, N.Y. U.S.A.
AXE
MADE IN U.S.A.

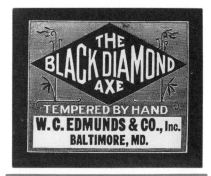

THE
BLACK DIAMOND
AXE
TEMPERED BY HAND
W. C. EDMUNDS & CO., Inc.
BALTIMORE, MD.

THE
WITHERELL
AXE
Copyright Secured
Preble & Robinson
Bingham
Me.

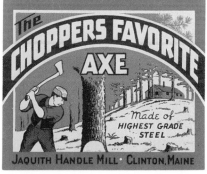

The
CHOPPERS FAVORITE
AXE
Made of
HIGHEST GRADE
STEEL
JAQUITH HANDLE MILL · CLINTON, MAINE

DAMON AXE
ALL HAMMERED
DAMON BROS. OAKLAND, MAINE

THE
LUMBERMAN'S
PRIDE
HAND MADE
WEDGE
AXE
Manufactured
only by
EMERSON & STEVENS MFG. CO.
OAKLAND, ME.

Should you ever stumble on a specimen like this old Collins, complete with sticker, I urge you to snatch it up (like Chris Garby did). Judging by the label, this one looks like it was made for the Latin American market

12. Notable (and Extinct) Axe Makers to Collect

Featuring mainly twentieth-century North American makers

Allan Hills Edge Tool Co. (Canada)

American Axe & Tool Co. (Pennsylvania)

American Fork & Hoe Co. (Ohio)

American Hdw & Supply Co. (Pennsylvania)

American Tool Co. (Kentucky)

Ashdown Hardware Co. (Canada)

Beatty Axe Markings (Pennsylvania)

Bedford Mfg. Co. (Canada)

Blood Axe Factory (New York)

Bradley Axe Co. (New Jersey)

Breckenridge Tool Co. (Ohio)

Canadian Foundries & Forgings (Canada)

Canadian-Warren A. & T. Co. (Canada)

Collins Co. (Connecticut)

Douglas Axe Co. (Massachusetts)

Dundas Axe Works (Canada)

Dunn Edge Tool Co. (Maine)

Emerson & Stevens (Maine)

Empire Tool Works (New York)

Francis Axe Co. (New York)

Geneva Tool Co. (Ohio)

Alexander Harrison (Connecticut)

Highland Tool Co. (Kentucky)

Hinds Co. (Vermont)

Hubbard & Blake (Maine)

Jamestown Axe Co. (New York)

Johnsonville Axe Mfg. Co. (New York)

Kelly Axe Mfg. Co. (Kentucky)

King Axe Company (Ohio)

Knickerbacker Axe Co. (New York)

Louisville Axe & Tool Co. (Kentucky)

Mann Edge Tool Co. (Pennsylvania)

Marsh Axe & Tool Co. (Maine)

M.F. Miller (New York)

Morris Axe & Tool Co. (New York)

Norlund Co. (Pennsylvania)

North Wayne Tool Co. (Maine)

Plumb Co. (Pennsylvania)

Rixford Mfg. Co. (Vermont)

Romer Axe Co. (New York)

Schreiber, Conchar, Westphal (Iowa)

Smart Mfg. Co. (Canada)

Spiller Axe & Tool Co. (Maine)

Tredway Hardware Co. (Iowa)

Walters Axe Co. (Canada)

Warnock Co. (Canada)

Warren Axe & Tool Co. (Pennsylvania)

Washoe Mfg. Co. (New York)

Wellan Vale Co. (Canada)

White (G.W.) Axe Co. (Pennsylvania)

Source: *Axe Makers of North America*,
by Allan Klenman, 1990.

Using

Michael makes camp in Andes, New York.

13. Safety

When you pick up an axe you bring something powerful to life. The tool is exclusively designed for productivity, but by its nature it can be destructive (no matter who's swinging it). As its handler you have a job to do, and that is to provide safety—not just to yourself, but to others, and to the world around you. Nothing is more imperative than that.

It's no coincidence that I'm starting the section on using axes with the chapter on safety and following that with sharpening. Before you use an axe you must make sure it's sharp, and if it's not you sharpen it, or you proceed with serious caution. The bit of the axe is where everything starts (and ends). A sharp axe is the surest way to guarantee that your axe will do what you want it to, and go where you want it to go. If you take the time to sharpen your axe, you will take your time to deploy it. A sharp axe says you mean to complete your task, and a dull axe says you don't really care. Never underestimate the damage a dull axe can do. A sharp axe is a safe axe.

A sharp axe is the safest axe

Each of the subsequent steps I outline in this section of the book will teach you how to properly use an axe, and they're written to maximize safety. But with all that, and no matter how hard you try, you will inevitably find that not everything is going to go as planned. No one has 100 percent control over an axe. No matter how seasoned you are, no matter how good or how sharp your axe is, there will always be forces you can't control. Your goal is to understand those forces and minimize their impact; that's the goal of this whole book.

In the wilderness, and far from medical attention, a lot can go wrong: Among many things you could break a bone, you could burn yourself, you could fall ill, or you could cut yourself with an axe. Of all the things that can go wrong, the last is most likely going to be the worst. A broken bone or a burn will hurt like hell, but you won't bleed from those injuries. Unless you're traveling with a surgeon or know how to suture a wound, simply stopping the flow of blood from a large wound, such as the sort that an axe can inflict, is incredibly difficult, if not impossible. It's the gruesome and harsh reality of using an axe, and something you need to imprint on your brain.

An axe injury is the worst kind

That said, what I love about using an axe is that it demands my attention. I have no choice but to focus, and everything else—all the nagging voices and distractions—has to take a back seat. The minute I lapse is the minute something goes wrong. There are certain times of day I prefer to chop wood: when I am most alert (and have a few coffees in me). I don't like to chop wood if I'm hungry and tired. If I'm chopping deep in my woodlot, away from the cabin, I will always bring water to keep me hydrated. Yes, that's "small advice," but the small stuff adds up to something big.

An axe demands attention

There are mistakes we make due to ignorance. There are mistakes we make by overcompensating (trying too hard *not* to make them). And then there are the mistakes we make when our guard is down and we assume we don't even have to try. I think these are invariably the worst mistakes. Consider this scenario: campfire at eight o'clock. You're tired from a long day of exertion, you've traded your sweaty work boots for your favorite camp moccasins, and you're four fingers into a bottle of your favorite scotch. The fire starts to die,

Keep your guard up

you're running low on firewood, and so without thinking you lazily reach for your axe . . . there is disaster written all over that scenario. Plan ahead and start with a raging woodpile before making a raging fire. Do all that in broad daylight, with your wits about you and your whisky corked.

A safe axe has integrity, but someday, and with enough repeated use, I guarantee you're going to break it. You will probably chip the head, but you're definitely going to snap your helve. It will break close to where it meets the head, a result of overstrike. When this happens, turn to page 171, buy a blank helve, and hang it on your head. You've just made a brand-new axe. Breaking an axe is not a big deal, and now that you know that, you can stop presuming your axe is indestructible.

An axe is not indestructible

With practice, you'll be able to put everything you've got into your swing, and your axe will absorb the blow without flinching, but why bother when a good axe will do the work for you? When you're chopping wood there's no need to rush, and there's no need to prove anything to anyone. Slow, deliberate, and methodical: that's what you need to be. A great chopper's power isn't defined by strength, but by an understanding of the axe, of the wood, and of the dynamics between the two. The great chopper takes that understanding and through practice hones his or her skill and builds valuable muscle memory. You don't have to be an expert to effectively use an axe, and with enough practice the motions will become second nature. As you learn, just keep in mind what you know and—maybe even more importantly—accept what you don't know.

Slow, deliberate, and methodical

The beauty of an axe is that it's going to take you outside and into some very special places. Always remember that there is a reason you are there, and that reason is not just to chop wood.

14. **Sharpening**

Most axe-sharpening guides I've come across consider sharpening a matter of maintenance, but I strongly differ. It's the critical first step in using. Likewise, most axe-sharpening guides I know have promised to show me the definitive way to sharpen a bit. But I don't think there is a definitive way. I think there are terrible ways to sharpen an axe, sure. But the definitive? Well, that depends on what you're looking for. Do you want your axe to have a mirror finish? That might take you days of work. When he's restoring an axe, my friend Nick Zdon might spend the better part of a week restoring and sharpening an axe, and he gets his axes so sharp and so shiny that you could shave with one and use another as the mirror. Is that mirror finish going to make it cut through wood better? Maybe a little. But it is a tell-tale sign that the axe is *sharp*.

Sharpening is the first step in using

The main thing is: Try not to let your everyday axe get so dull that you have to use a mill bastard file (one of the coarsest options) to get it where it needs to be. You remove a lot of material with an aggressive file like that. You're shortening the life of your axe. Sharpening is best done on a consistent basis with high-grit stones (see page 137). You'll get a better edge, and you'll be doing less work. Every time you sharpen it with a bastard file you're going to wear it down.

Sharpen frequently with high-grit stones

For regular upkeep, hone the axe before each use. You should be reaching for a high-grit Hard Arkansas stone, a natural product, or a high-grit diamond stone (600 grit or higher), a synthetic product. The best stone is the one you're most comfortable using. Synthetic stones are less likely to break, so they're good for use in the field. And some synthetics can be used without water or oil.

You'll feel more pride in your edge if you make the sharpening process your own. In the following pages, we'll show you how to sharpen a bit that needs some serious work. Give this a try. But use it as a guideline. I wouldn't deviate too much from these instructions, but sharpening, like the axe itself, is more nuanced than it appears. That's because no one really knows how sharp is the right level of sharp. Your "too sharp" may be my "not sharp enough." Now, axes can't be too sharp, but they can be too thin. A thin edge is a brittle edge. But if you go slow, you won't remove too much material too quickly.

Develop your own sharpening practice

Set aside time to sharpen—it shouldn't be an after-thought. Place everything you need in front of you before you begin. Your file. Your stone. Your cloth. Your bench vice (if it's not already attached to your table). Your axe. And clear the area of anything you don't need—includ-ing other people. Then move slowly but confidently until the job is done, picking up tools as needed and putting them back in the same place when you're not using them. The entire process should be marked by that level of care. It should be meditative, rhythmic. For two big reasons: Anything erratic makes for a less effective process. Smooth, efficient action ensures you're removing material in a con-sistent way, which will result in a uniform edge. But there's another reason to be so deliberate and slow: safety. Your fingers and thumbs and knuckles are going to be in close proximity with the very thing they're making sharper. Any fitful movements may lead to a cut. Deliberation and rhythm will mitigate the risk.

Sharpening should be meditative

Sharpening: Before you start

Before you start it's crucial to make sure you have all the right tools. I like to keep my sharpening bench clean and free of clutter, and typically all you'll find there is a bastard file, two or three stones (natural and synthetic), some C-clamps, some honing oil and linseed oil, a wedge, some rags, and a pair of gloves (and a cup of black coffee).

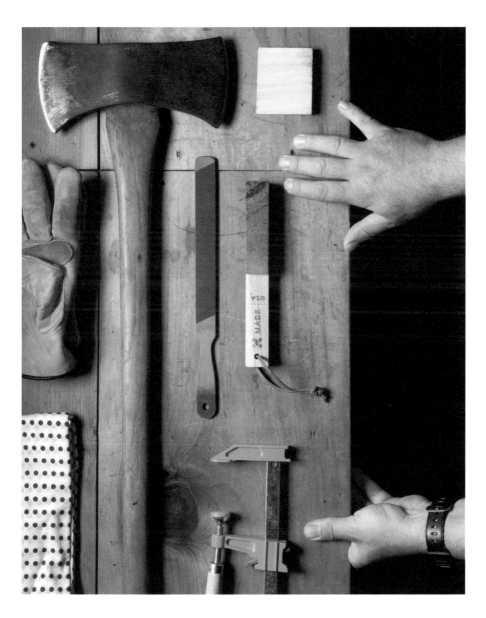

Step 1: Secure

Securing your axe to a sturdy work surface is important for both safety and efficiency. Use two bar clamps to prevent the axe from moving forward and backward (see page 183 to learn about a vice idea). Place the first about 8 inches (20 cm) below the head and the second farther down the helve (not shown). Use a rag to prevent marring the helve wood. The bit edge of the axe should overhang your work surface by a few inches.

Step 2: Shim

A wooden wedge placed between the axe-head and work surface will keep the axe-head from moving up and down and give your sharpening tools a better and more consistent angle of attack.

Step 3: Profile (top view)

For many restoration-project axes you'll want to profile the bit using a bastard file, especially if the axe is very dull. Start with the file at a right angle to the bit edge. Push the file into the bit and away from you. Never pull the file towards you. Move the file forward and then lift it off the bit completely, then reset. Work your way across the bit edge from one side to the next using slow, straight, consistent, and forward strokes.

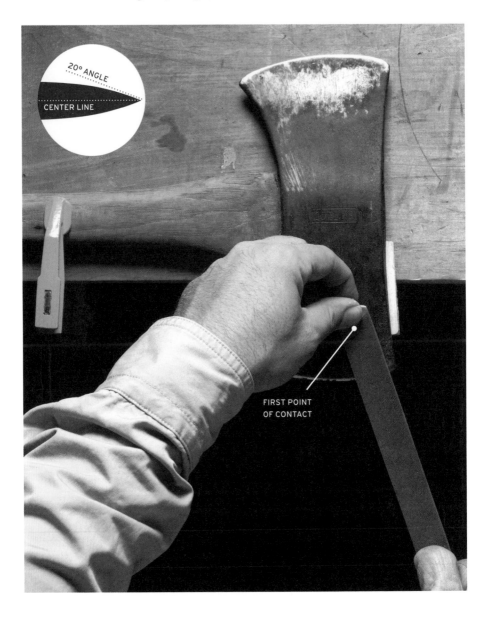

20° ANGLE

CENTER LINE

FIRST POINT
OF CONTACT

Step 3: Profile (side view)

Initial contact between the file and the bit should be 1 to 1½ inches (2.5 to 3.8 cm) back from the bit edge. As you push the file into the bit edge, increase the angle of the file using a subtle "scooping" motion. At the end of the stroke the file should be making contact with the bit edge. This motion is what creates the smooth, rounded, convex profile of the bit (see page 47). The higher the final angle the more rounded the bit profile.

Step 4: Check for the burr

As you continue to move back and forth across the bit with the file, a small amount of steel will be pushed to the underside of the bit. This is called a "burr" and is the natural result of your file stroke ending at the edge of the bit. Feel for it periodically while you're profiling and sharpening. Once you feel the burr all across the underside of the bit edge, you're ready to flip the axe over and repeat the process on the opposite side of the bit.

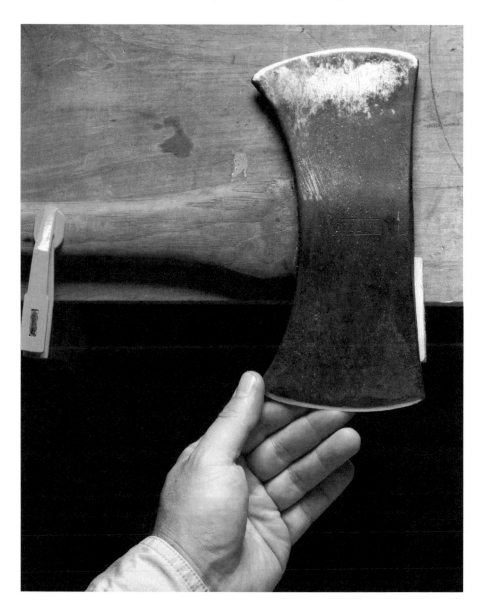

Step 5: Prepare your stone

Once you're pleased with your new bit profile and you've built a burr on both sides of the bit edge, it's time to start sharpening. From this point on you should be using a finer-grit sharpening stone (see page 137). First, prepare your sharpening tool. Arkansas stones work well with a liberal application of mineral oil, while diamond files work best with water. The liquid works to keep the stone unclogged.

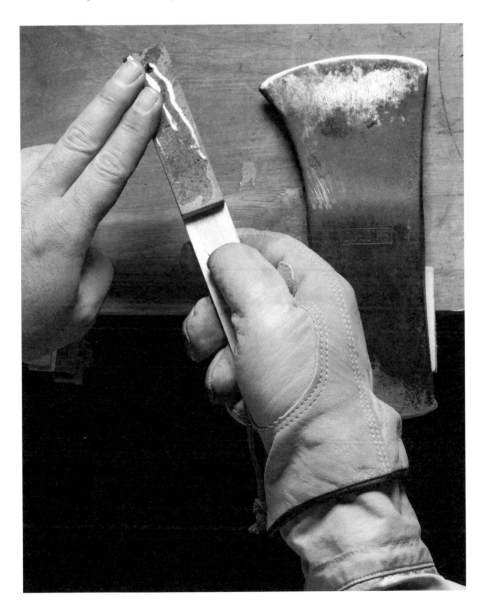

Step 6: Sharpen

The motions and angles when using a sharpening stone are the same as those when profiling. Push into the bit with enough pressure to feel the steel push back against the stone. Use slow, consistent strokes. Increase the angle of the file as you push, finishing at the bit edge. Work your way back and forth across the bit until you feel the burr develop on the underside (this burr will be more subtle than when using the bastard file.)

After you've sharpened both sides of the axe, you'll need to remove the burr that you've created. Do this by unclamping the axe and dragging the bit edge along a piece of cardboard or wood. Deburr one side of the bit edge, then flip the axe over and deburr again. It may take a couple flips back and forth to fully remove the burr. Once you no longer feel the burr on either side of the edge, you're done sharpening.

Sharpening in the field

If you're out in the field for an extended period you should assume you'll need to sharpen up. Always keep a portable round stone on your pack list. Not as effective as the bench method, a pocket stone is great in a pinch. Work the length of the bit in small, light circular motions, and repeat the same on both sides. Don't sweat small chips and nicks—they won't affect the axe's performance and can be fixed when you get home.

SHARPENING ABRASIONS AND GRITS

Numbers equal the grit (measured in sharp particles per square inch)

COARSE

FINE

FILE — For profiling and whipping extremely dull and/or chipped axes into shape. Should only be used sparingly as this removes a large amount of steel.

100 - 400 — For most axes, you'll start in this range. As with a file, just watch how much material you are removing.

500 - 1000 — You'll spend most of your time in this range. If your axe is sharp and needs a touch-up, you'll start with an 800 or 1000 grit stone.

1000 - 3000 — If you want to shave with your axe, and/or use it as a mirror, then you'll repeat the sharpening process progressively with these smoother stones.

PASTE — Often used with a honing strop. At this stage highly sharpened and polished bit edges are their own form of adornment.

STROP — Stropping polishes the edge and removes any final imperfections. Stropping will also bring your bit to a mirror finish. Use a leather strop (an old belt will do).

NATURAL VS SYNTHETIC STONES

The best stone for sharpening is as contentious and fiercely debated as almost any topic in the axe world. I use a mix of natural oil stones and synthetic diamond stones. Each has its advantage: I start with a natural stone and finish with diamond stones. I also like diamond stones for their convenience and portability. I suggest playing around with the stone that feels best to you. Sharpening is an art form in its own right.

15. Splitting

I love my hydraulic log splitter, but it comes with some offensive qualities: It's big and noisy and dirty. It dominates a scene that is inherently tranquil. The sound alone is enough to make me want to use an axe for splitting: the thwack, the splintering, the inevitable explosion that all somehow sounds as natural as a bird singing or a clap of thunder.

But you're not going to be able to consistently and cleanly tear through that wood—and build a magnificent woodpile—without being specific and intentional and following the same rules every time. Splitting wood requires practice and preparation, even if you've been doing it for years. It requires rhythm.

Practice, preparation, and rhythm

Speaking of rhythm: A maul or splitting axe is right for most splitting jobs, especially if you're splitting an entire "round" or cross section of wood. But I've always been partial to a felling axe. A splitting head is heavy—its mass pushes the wood apart more than it cuts into it—and its weight will tire you out. The felling axe is plenty heavy, and it bites deep; it just feels more secure in the hand and graceful in the air, and I've found it lends itself to rhythm more than its brutish cousins.

The maul vs. the felling axe

Although splitting seems easy enough—bringing a heavy blade down onto a piece of wood—it requires serious concentration and preparation. Don't be afraid to practice; go slowly, and take dry runs. The key is to build muscle memory, because once the axe is coming down, there's no going back.

The key is to build muscle memory

Preparation makes things go right. And so does time. Set aside time to split, and split in full daylight. And don't just go and go and go and whack and whack and whack. Splitting can easily tire you out. Be sure to take breaks and stop if you notice your attention slipping or that your axe isn't hitting its target.

I think of it as a kind of therapy. When I have guests up to my cabin, they eventually gravitate to the chop log and want to split. It's therapy for them, too. What is so gratifying about splitting wood? One, it's rhythmic. Two, it makes a great sound. Three, it's primal, just like hitting a ball with absolute precision. It's about doing your best to guide an object toward another, but there's a point at which things are out of your control. You can't predict with certainty what's going to happen next. And for a fleeting split second, you're in total suspense. Then the all-mighty explosion, the tearing apart of fibers that have been decades in the making, and from one single piece of wood you've now made two separate pieces, ready for the fire. As triumphant and easy a result as that can be, it could just as well end with your axe skipping off the log, straight into the dirt. Here's how to make sure that doesn't happen.

Splitting as therapy

Splitting: Before you start

Find the right chop log, something that's level and at least 18 to 24 inches (45.7 to 61 cm) wide. A good chop log can be your splitting base for years, so take good care of it. Make sure you clear the area of obstructions, people, pets, etc. Cut your logs straight; otherwise, you will have a hard time balancing them on the chop log. Save your gnarly knotty logs for the bonfire; I have broken and split many a helve trying to split the impossible.

Step 1: The right position

With arms fully extended, hold the axe with both hands at the butt end. Raise your axe so it hovers an inch or so above the target. Your feet should be spread wide apart (2–3 feet / 61–91½ cm) and your knees slightly bent—this will give you a stable foundation and also safe distance from the arc of a wayward axe. If in doubt, slowly lower the axe to simulate the arc of a wayward swing. If the axe touches your foot then adjust your position.

Step 2: Take aim

Slide your dominant hand up to the top of the helve and keep your other hand on the butt. Bring the axe in and across your body so the helve is almost touching your waist. With both hands in that position, raise the axe across your body and to the side of your dominant hand until the axe-head is above your head. At its peak height, the axe should be perfectly square over your body and stay that way through the rest of the swing.

Step 3: Raise the axe

With the axe at its peak height, you don't need to overly exert. Let the weight of the axe do the work; your job is to guide it safely home. As you bring the axe down to meet the target, slide your dominant hand down the length of the shaft to eventually meet your other hand on the butt end. Your wrists should not be locked. As you lower the axe, you will cock your wrists and the axe slightly up.

Step 4: Lower the axe

As you near the target you will snap your wrists and the axe down. That will deliver a "whip" action that will substantially increase force into the target. Chop down and through the chop log in a motion that keeps the entire axe parallel to the ground, with the business end of the axe terminating parallel with the top of the chop log. Avoid swinging in a circular arc, whereby the axe will terminate in the direction of your body.

Detail: The fist bump

You can see in these two images how your hands are the farthest apart as you raise the axe (below), and at its height they slide the length of the shaft and are together at the grip end of the axe as you bring the axe down (opposite). My friend Cristin Bailey, a trails manager for the US Forest Service, calls this the "fist bump."

Remember that whether you succeeded or not on your first try, the ultimate success when it comes to splitting is consistency. For that, you must find the position and motion that both works and is safe, build that muscle memory, and stick with it. Maintaining the correct position with absolute consistency will be your best means to ensuring repeated success.

Step 5: Contact

At the point of contact your body is at its lowest position; you are almost in a full squat with your knees bent, your back straight, and feet firmly planted. Ideally the axe will split the log cleanly. Don't be discouraged if you miss and the axe glances off your target: that's inevitable, and a chance to learn what's not working and adjust accordingly.

Step 6: Release and reposition

Hopefully you have chopped cleanly through the target and your axe is lodged in the chop log, and there are two freshly split pieces of wood to either side of you. Leave the axe in the chop log and clear the area for the next log. If you missed your target, just reposition the log and start again.

16. Limbing and Bucking

The first step in limbing, which is the removal of all limbs from the trunk of a fallen tree, is looking. If the bottom of the trunk sits high above the ground, notice which limbs are responsible for holding its weight. Know that the removal of these limbs will cause the tree to change position as soon as they're cut. With every severed branch, the tree's position will change. If you remove this limb here, it will roll a quarter-turn there. Removing that limb on the opposite side might mean it will roll right back. Removing that limb under there could mean the trunk slams to the ground. So, assess your targets. Figure out what the tree is going to do as you remove those large limbs. And make a plan for being clear of its path as it shifts around.

And then start limbing. Start from the bottom of the trunk and work your way up, chopping in the direction each limb is growing in. Don't start with any load-bearing branches. For thick limbs, make perpendicular notches, then come at them from an angle. Then clean up the area where the limb is attached. Your aim is a clean trunk that is as free of protrusions as possible. You want to be able to roll it easily when you're done.

Now, cutting the tree into large "sticks" so that it can be further processed into firewood, or bucking, is a much

Start from the bottom and work your way up.

bigger commitment than limbing. If you plan on bucking a lot of logs then I'd suggest you consider a saw. But there are definitely great reasons to buck with an axe. Axes don't take gasoline, don't require a partner to operate, and can go places that chain saws and large crosscut saws can't. Axes are great for work on remote trails, or when you have to traverse long distances on foot. Quite often, when I head out for a stroll on my property, I'll grab a light axe because I never know what might need clearing. (The drawback of using an axe to buck logs—other than the amount of time and energy you have to spend on it—is that it wastes a lot of wood. Think about the thin kerf of a saw versus the wide V-notch made by an axe.)

Why you'd
buck a log

Bucking can be dangerous work, and foot placement is important; I suggest keeping a wide and firm stance. When you're bucking you're chopping downward, at an angle. For big trees, you may choose to stand on the trunk you're chopping, and if so you'll be chopping toward one of your feet with every swing. The dangers are obvious. But it's worth doing—if only once. If you limb and buck a fallen tree all the way down to uniform pieces of firewood, you're connecting with the choppers long ago, the most able of whom (with just an axe) could buck, split, and stack two cords of wood a day.

Axe vs. saw for
bucking

(Note that a double-bit axe is perfect for limbing and bucking a tree. You can keep one bit a little thicker than the other and use it for cutting through the knot that forms the base of the limb, and then use the sharper felling edge for bucking.)

Limbing: The cut

Whether it's a tree you've just fallen or deadfall, the first step in processing a fallen tree is to remove the branches, or what's called limbing. When limbing a felled tree you want to start at the trunk and work your way up, chopping into the underside of each branch as you work your way up the length of the tree.

Limbing: The position

When you limb, always stand between the log and the branch you are chopping. Make sure your range of motion is always clear. Small branches are deceptively dangerous: Getting your axe caught on one mid-swing could spell disaster.

Limbing: One hand vs. two

You will always exert more control over your axe if you use two hands. When it comes to limbing trees, especially trees with a dense canopy, I will often switch between one hand and two. If you choose this method, then it's all the more reason to position yourself correctly, and watch for obstructions.

Bucking: Make your V-notch

Stand uphill from your target, out of the path of the log should it roll. Start by marking your V-notch. A V-notch should be two or three times wider than the log itself. If the notch is too narrow, your log will eventually bind and chips will become lodged in the cut. Have an escape route in case the log rolls.

Bucking: Angle your swing

Unlike splitting, when you buck a log you are coming in at left and right angles (varying about 35 to 45 degrees), with the goal of removing large chunks.

Bucking: Pace yourself

Bucking a log, especially a big one, can take a toll. Because you're swinging at an angle, there is a greater likelihood the axe will glance off the target. All the more reason to keep your axe sharp, and never rush. Be slow and deliberate with every swing. Stop if you fatigue and/or lose focus.

Bucking: The final blow

The moment we've all been waiting for. Be careful that you don't swing through the V-notch, sending your axe into the ground or back toward you. Also be aware of what your log will do once you've made that final blow: Double check if there are any branches holding it up. What direction will it roll? Once you release the tension, where will it spring?

17. A Few Words on Felling

I considered titling this chapter "Why You Should Never Fell a Tree with an Axe." There was nothing more hotly debated by me and my editors in the making of this book than including instructions on felling. I could teach you how to fell a tree, but I would only do that in person and in the field, not in the pages of this book. I don't suggest you try and learn how to fell a tree by watching a video (and there are plenty of them out there). If you're going to fell a tree, and you have good reason to do so, please don't grab an axe. Grab a saw—preferably a crosscut saw or a chain saw. And then put the saw down, grab a phone and call a professional.

Think before you fell

Of all the things you can do with an axe, felling a tree is by far the most dangerous. If something goes wrong felling a tree, the result can be death or severe and lasting bodily harm. I have a friend who was paralyzed from the waist down by a tree he didn't know how to fell.

There are an infinite number of variables that go into felling, but let me just simplify a few. The weight of a 16-inch- (40.6-cm-) diameter hardwood tree—which you would probably consider a relatively small tree—is approximately 1.5 tons, which is approximately the weight of a small car. Imagine what a small car would do to you if it was dropped 15, 20, 30 feet (4.6–9 m) from the air. When

The reality of felling

you fell a tree you're manually releasing an exceptional amount of force. You are standing within inches of that exceptional force, and once it's released there is no going back, no undo button: Once you make the last cut, that tree may go where you least expect it.

Axes were once the only tool able to get the job done—that's why they're called "felling axes"—but now there are better tools. It's important to note that axes are still a vital part of the felling process. Even if you use a chain saw, you will still need an axe to pound in wedges or to free up a bound saw bar.

There are better tools to fell with

Why would you need to fell a tree when ninety-nine times out of one hundred there will be deadfall or windfall to limb and buck? If you need firewood, and lots of it, then you may need to know how to fell a tree. If that's the case then I suggest contacting your local forester and asking the best place to learn how to fell a tree, most likely with a chain saw. I am a fan (and a student) of Soren Eriksson's course "Game of Logging," a chain saw class taught in various locations in the United States and Scandinavia. I believe it instills the right techniques and safety procedures for felling.

So you want to fell

I hope I have discouraged you from ever wanting to fell a tree. That said, I have to admit that once you know what you're doing, felling is one of the most rewarding and thrilling things you can do with a saw (not an axe). Good woodlot management depends on the removal of both dead and live trees. If your desire is to harvest wood for firewood or other purposes, then start with deadfall—trees that have fallen by natural forces. Then work your way to trees that are inhibiting the growth of other trees, trees that possibly pose a safety threat. And by that I mean call in a professional to do it for you.

REASONS TO FELL A TREE

1   You know what you're doing.

2   You have a crosscut saw or chain saw.

3   You have a serious need for a lot of wood.

4   I repeat: a serious need.

REASONS TO NOT FEEL A TREE

1   You don't know what you're doing.

2   You only have an axe.

3   You have no real need for a lot of wood.

4   There's deadfall available.

5   The tree is big enough to kill you.

18. Small Axe Work

At the time of publication of this book, there has been a major proliferation of small-axe use (as the founder of Best Made I sold more hatchets than I can count). But as useful as a small axe is, it's deceptively dangerous.

Your first axe may likely be a small one, probably purchased for a camping trip, or as an all-purpose utility tool to keep around the house, good for kindling or felling the next Christmas tree. Throughout its life your small axe will certainly come in handy, but it will go relatively unnoticed, and that's the problem: There are hundreds of thousands (if not millions) of small axes floating around the world, owned by thousands of people who don't know how to use them safely.

The danger of a small axe

Keep your small axe in a safe place, and treat it with the respect you would a large axe. Keep it out of reach of children and friends and family members. If you travel with your small axe then make sure it's kept in a safe place, securely stowed and lashed down. When it comes time to deploying your small axe, the same basic rules apply as they do to a big axe: Make sure your axe is sharp, that the area is clear, and that you have full visibility.

Same rules apply to small axes as they do big

You will most likely use a small axe with one hand, and that automatically means your other hand has nothing to do, and it will want to "lend a hand." Invite your spare hand to help prepare your target area, but when it comes time to swinging, keep it well away or use two hands to chop.

Don't let the relative "cuteness" of a small axe fool you: They can bite. If you miss with a big axe, it will likely glance off the target and land away from your body (and probably in the ground). When you miss with a small axe, the chances of it landing in your body (your shin, your hand, your femur, your knee) are much greater. If I'm in a remote location, far from medical attention, I will not only use a small axe with two hands, but in some cases I will kneel. Kneeling removes the better part of your legs from the equation.

Obviously, small axes are great at doing small work, like shaving wood off a larger log to make kindling. A small axe like a Hudson Bay is designed with a cutaway head that allows you to choke up on the helve, get your hand close to the head, and thus gain more control of the axe.

On his remote island in Patagonia, I witnessed the legendary Argentine chef Francis Mallmann use a Hudson Bay axe to make short order of his famous lamb asado. He gripped it tight with one hand and made discrete, efficient cuts. It was probably the most controlled use of a small axe I've ever witnessed. And the fact that he was using it to cut meat, not wood, brings up an important point: The small axe is versatile. But no matter how you're using it, the same rules always apply.

SAFETY TIPS FOR SMALL AXE WORK

Take a Knee

Choke Up

Use Two Hands

Cover the Blade

Drilling the old helve out of the eye to make way for the new one.

Maintaining

19. Hanging

Hanging an axe is the brave and noble pursuit of attaching
(hanging) an axe-head to a helve (not on a wall). An axe-
head on its own is a heavy, albeit sharp, and unwieldy piece
of steel. A helve is a long piece of wood, and on its own,
it's even more useless than an axe-head. You put the two
together and it's a small miracle. This is what I love about
hanging: You are essentially making your own axe. You are
a miracle worker.

Hanging an axe
is making an
axe

Few tools can endure time as elegantly as those made of
wood and steel, and the axe is particularly egalitarian in its
method of repair. Axe hanging can be accomplished with
a few basic hand tools and requires more patience than
expertise. The result of renewing a well-worn helve with a
fresh length of hickory not only extends the axe's utility, but
leads to prideful ownership as well.

Hanging is a tradition worth maintaining, in support
of sustainable maintenance over disposable replacement,
and it allows retired bits to be returned to service for at least
another generation. The process of hanging a helve makes
the benefit of a wood helve over fiberglass explicitly clear.
You can't argue that shaping a fine specimen of wood to
perfectly fit a unique piece of steel is much more gratifying
than sealing a fiberglass helve with two-part epoxy. Not

Sustainable
maintenance
over disposable
replacement

only is the process of fitting a hickory helve less toxic, the helve itself can be tossed into the campfire (or your smoker) if it ever needs replacing—the effect would be a lot less pleasant with a fiberglass helve.

It almost seems impossible. How can you join wood to steel in such a way that the head won't fly off the five hundredth time you swing it? Or the first? How can a single person with only a few tools drive a piece of wood into the eye like that so that it fits securely? How much brute force does that require? The secret is that the process isn't about brute force—or glue. It's actually a fairly gentle process. You're almost coaxing it on. It requires patience and just the right amount of pressure.

Patience over brute force

Like sharpening, the secret to hanging starts with patience and taking your time. If you think you're going to get it the first time round, if you expect the first axe you hang is going to last forever, then you might as well not bother. Getting good requires dedication, but it doesn't require any elaborate woodworking skills or tools. Hanging is just about removing wood and shaping the helve to fit perfectly into the eye of the axe and then gently coaxing it in. Progress, especially to start, is made in small increments, whereby you may be fitting and refitting the head back and forth onto the helve multiple times until it's just right. The better you get, the fewer the increments. The final step will be inserting the wedge, but don't assume that will make up for previous missteps or corners cut. Remember: The axe functions as one continuous system, the more seamless the union of wood and steel the better. The more breaches in that system, the more problems you're going to have down the road.

The axe is a continuous system

Hanging: Before you start

Like sharpening, start with the right tools. Mine include some rasps, a mallet, a bench vise, a drill or drill press, and a saw (a coping saw, a flexible Japanese flush-cut saw, or a band saw). As far as supplies go, you'll need some spare wedges, wood glue, and linseed oil. Most importantly, you'll need a replacement helve: There are good resources for those online, and if you're so inclined you can carve your own.

Step 1: Decapitate

Most axes that need hanging are going to be the result of a broken helve, and the first order of business is to extract the old helve to make room for the new one. For this you'll need a saw (here we're using a band saw). Note: 99 times out of 100 it's a waste of time, and dangerous, trying to repair a broken or cracked helve, and better to just hang a brand-new one.

Step 2: Drill it out

It can be deceptively difficult to remove an old helve from a head; some of the older ones can seem permanently fused in there. Use a drill to remove the majority of material inside the eye of the axe-head, but just watch for steel wedges, nails, screws, and other inventive flotsam that's been jammed in over the years to keep the head attached. Whatever you do, don't burn the helve out: you'll probably ruin the temper of the steel.

Step 3: Fit the head

By far, the most tedious and time-consuming step in hanging an axe is getting the helve to fit just right in the eye of the axe. The eye of an axe-head is not a perfectly straight cylinder; it is likely tapered and requires repeated fittings and refittings. To start, if the axe-head slides right down on the helve, then you know the helve is too small. You should have to remove wood with your rasp to even partially seat the head on the helve.

Detail: Support the helve

As you hang an axe in your vise, you will be tapping the head with your mallet and all the force will be applied downward. When your axe is secure in the vise, there will likely be some space between it and the floor. It's critical that you shore the axe up so you have some resistance (other than the vise). In this case we recruited a small block of scrap wood.

Step 4: File and fit

See the the black marks on the side of the helve in the photo below? Those are the marks made by the inside of the eye of the axe-head as we seated it (tapped it down with our mallet), and these marks are crucial indicators: They show where the head is making contact with the helve. Using your rasp, you'll file these marks away. Once removed, you'll reseat the head, and it should seat farther and farther down the helve.

Step 5: Invert hang

Once the head is close to being seated on the shoulder of the helve, you'll remove the axe from your vise (with head firmly seated) and flip the axe upside down. Using your mallet, you'll pound the butt of the helve down, but with each blow your axe-head will defy gravity and magically creep up. This step will seat the head far more securely and effectively than pounding the head in from the top.

Step 6: Fit the wedge

The final piece of the puzzle is the wedge. The goal is to make sure the wedge fills as much of the void of the kerf slot (the opening at the top of the helve) as possible. To start, cut the ends of the wedge down until it fits perfectly (widthwise) into the kerf slot. With the head perfectly seated on the shoulder of the helve, you will insert the wedge by hand into the kerf slot.

Step 7: Insert the wedge

The entire wedge doesn't have to fit into the kerf slot, but the more the better. If your wedge is coming up short, then it means it won't effectively bind the helve to the head, and you will need to file away more helve to make room for the wedge. Once the wedge fills the void of the kerf then you'll pound the wedge down with your mallet. To avoid snapping your wedge, only strike it with your mallet parallel to the axe.

Step 8: Remove excess wedge

When the wedge will go no further, it's time to remove the excess. Some people like to keep a good inch of excess wedge, under the assumption that after a few hours of chopping the head may have a little play and they have more wedge to pound down and resecure the head. After the excess is trimmed, be sure to apply a liberal amount of linseed oil to the exposed grain in the eye, or soak it in linseed oil overnight.

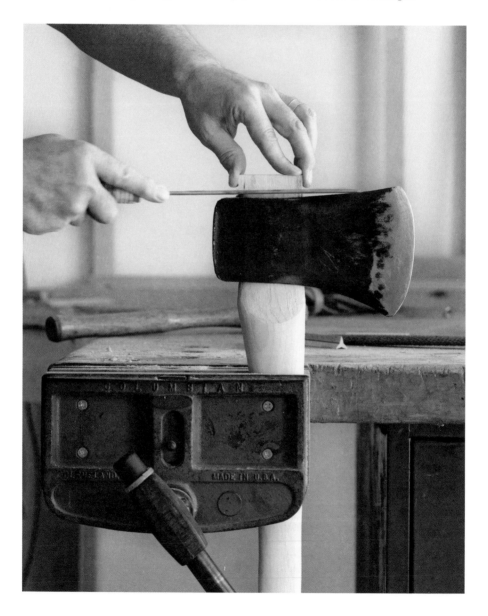

THE DUDLEY AXE VICE

Attempting to sharpen an axe that's improperly secured to the workbench
will turn gratifying maintenance into pure drudgery. The simplest
method is to clamp the helve at two places along the workbench edge
and slip a spare helve wedge underneath the bit to support the cheek.
This configuration prevents rocking under the pressure of the sharpening
stroke and is fairly quick to switch between sides of the edge. Some folks
simply clamp the axe bit in a bench vise, which is certainly the quickest to
reposition. We've found that the horizontal sharpening stroke is much more
natural, however, and allows for a better view of the sharpening process.

Peter Dudley, the expert woodworker and ex-master of the Best Made
Axe Shop, has devised an ingenious workshop jig that works better than
anything else I've tried. A simple leg vise with a wing nut is mounted
horizontally and registers against the edge of the bench. Both bit and
helve are held at a comfortable angle on padded supports. Two vertical
dowels provide consistent placement when readjusting. An elegantly
simple jig that has been refined with practice, the Dudley Axe Vice can
be easily replicated in any workshop.

Before

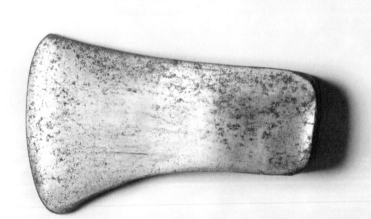

After

20. Restoration

Restoration is as important a skill as any in this book. It resurrects a dead tool. It furthers its legacy. And it (quite cheaply) gives you a brand-new axe. However, it's a subject not without controversy. Should you use harsh acids like straight vinegar to remove rust? Should you take the axe back to its original state (or even make it "better")? Or should you try to preserve its history, its scars? This same debate rages in the worlds of art and architecture.

There are axe restorers for whom restoration means preserving the axe in whatever state it's in, and there are those who look to change the entire profile of the axe—removing the beard as you would extra material from a sack suit. I'm somewhere in between, and so is Nick Zdon, who helped me start the Best Made axe-restoration program. You should acknowledge such controversy. You should also acknowledge that axe restoration is, quite simply, fun and rewarding.

How much do you want to restore?

The hardest part is deciding what axe is a prime candidate for restoration (see page 105 on buying old axes). Everybody has their own criteria. Say you find an attractive but rusty old, anonymous axe on eBay—something that caught your eye while scrolling and isn't too pricey— then why not strip it down by soaking it in vinegar? But if you're restoring your grandmother's axe (what a woman), or any axe that tells a story worth passing down, and you

Chosing the right axe to restore

want to keep that patina on the head and the pitch on the wood, then steel wool or an angle grinder with a brass wire wheel is probably the way to go. Patina is itself a kind of preservative—of both material and myth.

My friend Nick, who's restored hundreds of axes in his own shop and during our time together at Best Made, looks for axes that haven't gotten a lot of use or abuse. Only light mushrooming on the poll due to pounding (again, a poll

Rust is not as bad as you think

is a counterweight, not a hammer) or maybe a corner of the bit has been snapped off. If you see those problems but the head has a good shape, you've got a good candidate for restoration. You should like the axe, oxidation and all. Rust is not a big deal. You can remove it. The condition of the helve is not a big deal. You can rehang. You simply need a good head shape, something with crisp clean lines, something that hasn't been over-ground or over-"restored" or terribly abused.

About the vinegar. The problem is that it can actually remove metal. It can take it back to a gunmetal gray and sometimes even reveals the different metals in the head. But several coats of oil usually solves that problem. Some restorers build their own electrolysis machines to remove rust. Or you can just use a grinder to remove rust. Which, if you do it right, can be the gentlest method. And if you do it wrong, it can be the harshest.

Which brings up a point. The axe isn't like a hand plane, which has been precisely machined. No, the axe was

Restoration is an experiment

forged in such a way that metal can be removed over many, many years. It's okay to experiment. And it's okay to screw up. While this book is imbued with a certain reverence for this tool, it's no tragedy if you go overboard restoring your axe. The skill and edification you receive in the process will make future restorations go more smoothly. And thus why I give myself serious price constraints when buying old axes.

CHEMICAL VS. ABRASIVE AXE RESTORATION

When it comes to removing rust from a neglected axe-head, there are
generally two camps: chemical and abrasive.

CHEMICALS

WHITE TABLE
VINEGAR

APPLE CIDER
VINEGAR

BAKING SODA AND
LEMON JUICE

CITRUS-BASED
SOLVENTS

NONTOXIC RUST
REMOVER

ACIDIC SOFT
DRINKS

The chemical method uses a weak acid to dissolve rust.
These acids react with ferrous oxide, forming a salt that
washes away from the metal surface. Acids are corrosive,
however, and will begin to eat away at the clean surface of
exposed steel if left unattended. There's a sweet spot where
the rust has been removed from the surface and the acid
has been given just enough time to form its own oxidized
patina, making the steel more rust-resistant than if left
bare. But hitting that sweet spot takes a watchful eye, so
axe-heads should never be left in acidic solution without
keeping an eye on things. As expected, the chemical method
is a "fun with science" experiment. Take apple cider vinegar
for example: Soaking an axe in that versus, say, a white
vinegar produces an entirely different result (I'll let you
undertake that experiment and see for yourself). When you
are finished, dispose of your spent vinegar down the drain.

ABRASIVES

ELECTRIC GRINDER

BELT SANDER

ORBITAL SANDER

WIRE BRUSH

SANDPAPER

The second camp of rust removal is the abrasive method,
which involves removing rust without chemicals and the use
of mechanical and hand tools. The least aggressive tool is
a brass-wire brush (make sure the bristles are true brass,
not brass-coated wire), which is softer than steel and will
not mar the surface. A hand-held brass wire brush is safe
in just about all circumstances, but for this application a
drill-mounted or bench-grinder wire brush is the right move.
The wire brush will remove the rust with minimal impact on
the steel face—this preserves any characteristic pitting of
the forging process or age. (I've always loved that pitting.)
Be careful with this method, especially around the bit. If
the abrasive action heats up the steel too much, it can ruin
the temper. Use any electric abrasive device at its slowest
setting and start with a low grit. To bring out the character
of a well-aged axe bit, use a wire brush on the sides of the
bit, and only use sandpaper to clean up around the edge.

Step 1: The vinegar bath

Fill a shallow tray (deeper than the axe-head) with a vinegar of your choice and insert your axe-head. Leave the head soaking in the vinegar undisturbed for at least 12 hours, and no more than 24 hours. After 12 hours, feel free to run a finger over the head; if not much rust comes off then keep it in there.

Step 2: The first scrub

Using latex gloves, remove the head from the vinegar bath. Without gloves, the vinegar-rust solution won't hurt your hands, but it will stain them. Next, gently scrub the axe-head with a household cleaning device (steel wool, a scrub sponge, an SOS pad). Be sure to scrub all surfaces of the head, and even inside the eye.

Step 3: Rinse and scrub

After a few minutes of scrubbing, your head should start to reveal itself, and the rust should be easily wiped away with a clean paper towel. Wipe the head clean and continue to scour. You can see (below) there are some old signs of blue paint on this head: Feel free to keep or remove any paint, patina, or pitch below the surface of the rust.

Step 4: And repeat

If the rust is especially stubborn, then you may need to soak your head a little longer in the vinegar bath. Likewise, you could also choose to use more aggressive abrasives. Just watch the bit of the axe and how much material you remove there.

Step 5: Sand

Ideally, as you restore your axe-head your abrasives should be getting finer and finer. Below, I am taking an 800 grit wet/dry sandpaper to the head. How far you want to restore all depends on how much material you want to remove. With enough time (and sandpaper) you could bring your axe to a mirror finish.

Step 6: The finishing touch

As you can see below, I like leaving many of the marks and traces of this axe's former glory. With this head almost finished, all that's left is to give it a coat of mineral oil (to protect it from rusting again) and then it will be ready to hang on a new helve and sharpen. Dry thoroughly to prevent flash rust, and coat with oil immediately.

21. Adorning

It's hard for an axe to be ugly. Even the least sentimental axe wielder has to admit: The combination of head and helve creates a tool that's visually appealing in a way many other pieces of gear simply aren't. Even a terrible axe is nice to look at. Even one with a head pattern you're not especially fond of. Even one that's been rusting in a barn for a hundred years.

And you can always make an axe better. I started a business by doing just that. I painted that first Best Made axe because I loved the tool; I wanted to pay homage to it and wanted people to love it. I subsequently sold thousands *A blank canvas* of adorned axes, shipped them the world over, and built a thriving business around them. I know the wonderful effect of a brightly painted axe. And while running Best Made, I received countless notes and photographs from proud customers who had taken it upon themselves to adorn their axe (or their grandfather's axe). The end of an axe helve is a blank canvas if there ever was one.

You can paint a helve with bright colors, carve poetry into it, scorch it. But you're not just making it more beautiful.

Adornment is an act of art, sure, but it's also an act of respect for the tool and the skill it takes to use it. Furthermore, it imbues your axe with an individuality—one it possessed the moment the helve was hung, even if it looked like every other axe created that day.

The license to love a tool

That act will be part of a rich tradition, no different than engravings on the barrel of a revolver or shark teeth on the nose of a P-40 fighter plane. Human beings have adorned their tools for as long as they've invented them. I think it's because adornment gives us the license to love the inherent beauty in an object of utility.

Which is a powerful feeling—and one axe manufacturers have exploited since the first axe was sold. During the golden age of the axe, which lasted from the mid-nineteenth century to the early twentieth century, adornment was a marketing tool. Manufacturers applied labels directly onto the cheek, invariably with the name of the product, the name of the manufacturer, and in what town the head was forged, all of which came to life set among an idyllic nature scene, at the feet of a giant redwood, the mouth of a river logjam, or set into an impossibly elaborate typographic engraving. If you're incredibly lucky, you can walk into an antique store and find the remains of these stickers on axes now. If you see one intact, you know the axe was rarely, if ever, used.

Know what's underneath

Adornment can also have the effect of covering up inadequacy. Some axes come with heads dipped in paint, which covers up a lot of the evidence of how the axe was created: the forge marks on the head—or lack thereof—as well as breaches in the eye of the axe, the wedge, or the kerf. Gränsfors Bruks axes used to come with heads painted blue. Among other things, legendary axe maker Gabriel Branby eliminated the paint when he took over the company in 1982, and had each blacksmith stamp their initials into the axe. A Gränsfors blacksmith had nothing to hide behind, and quality soared.

The next time you see footage of wildland firefighters in the American West, or firefighters on the streets of New York, note the axes they're wielding. If you look close, you'll often see a set of stripes emblazoned on the helve, up near the head, denoting which crew or ladder company the axe belongs to. For these firefighters, who literally live and die by the tool, these are symbols of solidarity.

One of the most impressive axes I ever laid eyes on was one with a simple phrase crudely brandished into its helve: "THIS AIN'T YOURS." That priceless gem belonged to a mule packer I ran into out in Idaho's Frank Church Wilderness. It doesn't belong to anyone else, and it never will. He knows it. Anyone who came within spitting distance of that axe knew it. It was axe poetry, and I'd nominate that packer for a Pulitzer. Regretfully, I never had the chance to ask the packer what the name of his axe was. But I bet it had one. Anything that near and dear should have a name.

Like restoration, one of the best things about adorning an axe is the reward. I started Best Made in my small garage (see next page) with no more than $100 worth of supplies. The earliest axes I sold were finished very much as Best Made axes are today: gessoed, sanded, meticulously taped off, painted with a spray enamel, and finished with a few coats of marine spar varnish. I loved handling each axe and elevating this amazing tool with every layer of paint. I found a real satisfaction in perfecting my process. I loved to experiment with new colors and new patterns. And the beauty of it all was that if it didn't look right, I could just sand it off and start over. No matter how easy or complicated you make it, nothing beats taking the tape off after that last coat of varnish and beholding your creation.

The reward

THE BEST MADE WAY OF ADORNMENT

This is a photo taken in my workshop in the spring of 2009. These were the first eleven axes I ever painted—they were my perch to the brand I had just founded called Best Made Company. To the right is a poster that I made shortly after, an array of abstract swatches that advertised some of those earliest axes. I only ever painted the bottom third of the helves. What can I say? That space—and nothing more—just felt like it needed a coat of paint. I used bold, graphic colors and patterns that mimicked nautical and military vernacular. I had been obsessed with polka dots for a long time, and sometimes I tell people that I started Best Made just to make a polka-dot axe. Before any axe left my workshop I gave it a name. "Famous Red" because I wanted to corner the market on the color red. "Pale Male" after the famous red-tailed hawk in Central Park. "Flashman" after a childhood storybook hero. "Auld Reekie" after one of my favorite cities (Edinburgh). And "Conacher," an homage to another childhood hero (see page 1). I knew that axes could tell stories, they just needed a name to get them going, and the rest would soon be given by my customers.

OUR FAMOUS FOUR

GRACE

SUMMER 2009

SPRING 2009

✕ BEST MADE

368 BROADWAY #514
NEW YORK, NY
10013

BESTMADECO.COM

FAMOUS YELLOW

FAMOUS BLUE

FAMOUS PURPLE

FAMOUS GREEN

CONACHER

HI-LO

ZEPHYR

FLY-HALF

OLD RADIANT BEAUTY

ROYAL STANDARD

PALE MALE

AULD REEKIE

BIG SUR

MOSS THUMPER

FAMOUS RED

FIRTH OF FORTH

JIM DANDY

FAWN'S FEATHER

PIOBAIREACHD OF DONALD DUBH

HANDSOME DAN

(PICTURED ON FRONT)

CANUCK CLIPPER

DREADNAUGHT

DILEAS GU BAS

FLASHMAN

Step 1: Sand

You'll need a fresh start with most axes, even new ones. Using an orbital sander or sandpaper, remove any existing varnish, dirt, and markings. Start with coarser-grain sandpaper (400) and work your way up to increasingly finer grain (1200). If you are using an orbital sander, then be sure to keep the sander moving in long strokes, and beware of removing too much material. Sanding is also a chance to shape the helve.

Step 2: Tape

Using blue painter's tape, determine where you want to make your mark: This could be as simple as one solitary band, or an intricate array that will make a plaid. Use drafting dot stickers to make a polka-dot pattern. I personally prefer painting the bottom third of the helve and leaving the rest blank. The helve can get too grippy if it's entirely painted.

Step 3: Gesso

Using a sponge brush, apply two to three coats of gesso (the more coats the smoother the surface). Gesso is basically an artist's primer, normally used to prime canvases. Sand by hand with a fine-grit sandpaper (1200) between coats. Let dry per instructions on gesso package.

Step 4: Mask

Using newspaper or scrap paper, you will mask off the exposed (un-gessoed) part of the helve. Spray paint has a tendency to find its way into any unexposed crack or crevice, so it's important to protect any surfaces that you don't want to paint.

Step 5: Spray

Make sure your work area is well-ventilated. There are small portable vents with replaceable filters that will extract the majority of fumes; however, if you are in doubt then take it outside. Protect your work area thoroughly with newsprint or drop cloth. Wearing a respirator mask that is rated for enamel spray paint, apply a series of thin coats of paint. Take your time and watch for drips.

Step 6: Varnish

Once the spray paint has fully dried, use a sponge brush to apply multiple coats of varnish. I use a marine spar varnish because it's one of the toughest coatings and adds a shiny look. I've also used polyurethane coatings, which will offer multiple types of finishes. (Note: Avoid varnishing the entire helve—the wood needs to breathe. It's OK to varnish the lower third. This protects the paint and will give you needed grip.)

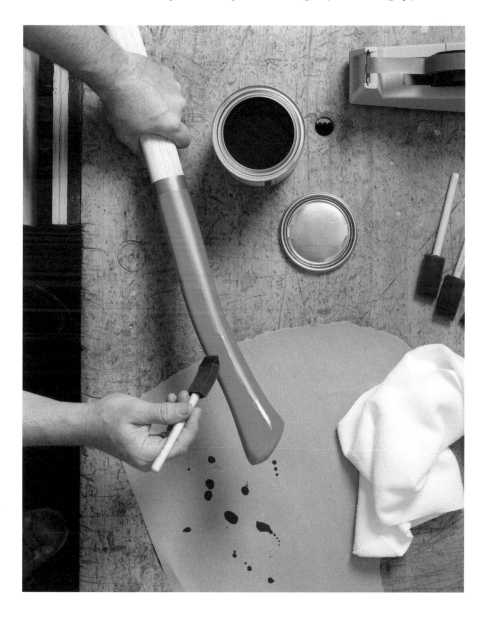

Step 7: Oil

The last step is to oil all the exposed (unpainted) parts of your helve. Using a sponge brush, apply two to three coats of boiled linseed oil to the helve and the exposed end grain in the eye of the axe-head. Wipe excess linseed oil down with a rag or paper towel, but do not dispose in the garbage. IMPORTANT: Any rag, paper towel, or cloth that has come in contact with linseed oil should be disposed of in an appropriate oily waste can designed for flammable materials, or hung outside to dry.

THE JAPANESE WAY OF ADORNMENT

Harnessing the ancient Japanese technique called Shu Sugi Ban (SSB),
Peter Dudley uses a propane torch to scorch the end of a felling axe. This
technique hardens and purifies the wood while giving it an incredibly
deep-black pigmentation. After you've scorched your helve to the desired
amount, wipe with a rag and apply a coat of varnish or linseed oil. Unlike
a painted axe, an axe adorned in the SSB style will never chip or discolor.

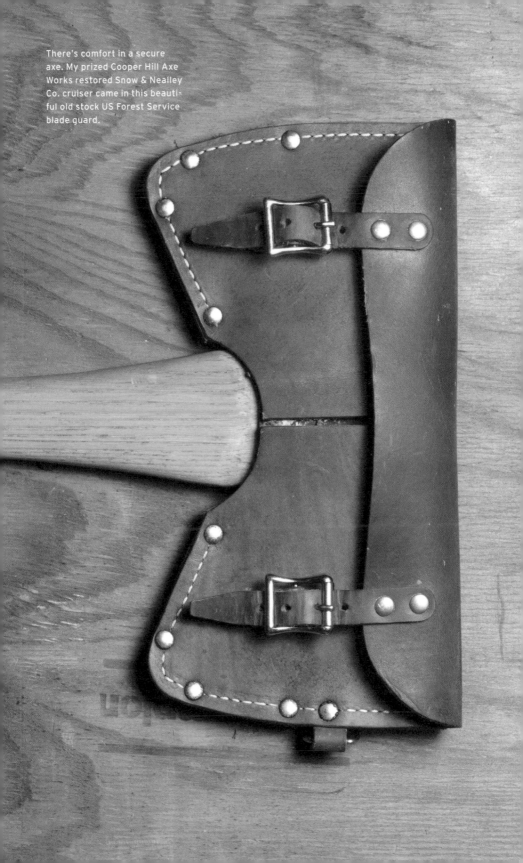

There's comfort in a secure axe. My prized Cooper Hill Axe Works restored Snow & Nealley Co. cruiser came in this beautiful old stock US Forest Service blade guard.

22. Storing, Handling, Upkeep

While leaving an axe stuck in the chop log makes for a pretty picture, it's a terrible way to store an axe. For short-term storage, the bit should be wiped clean and lightly oiled. Mineral oil works best, though any nonpolymerizing oil that doesn't leave a residue can be used. If the helve feels dry, rub it with boiled linseed oil (see page 206 for rag disposal instructions). The rule of thumb is to oil your helve once a day for a week, once a week for a month, and once a month for a year. The helve will be plenty saturated at that point and can be oiled as needed. In the short term, the axe can be stored in a sheath. But only for the short term. Leather, especially the rough-grain interior of most sheaths, is excellent at trapping moisture in its pores, and will be sure to degrade the steel over time.

For the long term, if the chopping axe is put away for the summer or another axe has taken priority, both helve and head should be oiled thoroughly. I've heard of folks who prefer a more viscous oil, such as a summer-weight motor oil. You can even use a wax product to protect the steel from moisture. For year-to-year storage, a coat of mineral oil works fine, as long as the axe is kept where moisture isn't allowed to sit. Hanging in a well-ventilated shed does the trick. You just have to make sure the axe is stowed in an extremely safe place.

Transporting the axe is straight-forward: If you protect the blade, then you protect yourself. When carrying in hand, hold the axe on its shoulder, just under the head. This grip gives you the most control over the business end. Slinging an axe over your shoulder might look cool, but if an errant root grabs a hold of your boot . . . well, that wouldn't be cool at all. If careful practice is observed, the axe can be carried unsheathed while cruising for timber (if you're a lumberjack) or looking for wood (if you're not a lumberjack). If you have any doubts about safety, there's little reason beyond inconvenience to put the sheath back on. If the axe is lashed onto a pack, be sure to secure it at three points: the shoulder, the grip end, and the bit. This arrangement prevents the helve from kicking outward away from the pack, and keeps the bit from rotating away from the pack. When transporting with other gear, a canvas bag can keep the axe from experiencing unnecessary wear, or the axe can be lashed to a larger piece of gear such as a roof rack or saddle.

Index

Resources, Credits, Locations

BOOKS, CATALOGS, PAMPHLETS:

The American Axe and Tool Co. Catalog, 1894

The Ax Book: The Lore and Science of the Woodcutter (formerly published as *Keeping Warm With an Ax*, 1981), by D. Cook, 1999 (Alan C. Hood & Company)

Axe Manual, Peter McLaren, 1929 (Fayette R. Plumb)

Camping in the Old Style, David Westcott, 2009 (Gibbs-Smith)

Axes, Oxen, & Men, by Lawrence C. Walker, 1975 (The Angelina Free Press)

Axe Makers of North America, by Allen Klenman, 1990 (Whistle Punk Books)

Yankee Loggers: A Recollection of Woodsmen, Cooks, and River Drivers, by Stewart H. Holbrook, 1961 (International Paper Company)

Northeastern Loggers' Handbook, 1951 (US Department of Agriculture)

The Axe and Man, by Charles A. Heavrin, 1998 (The Astragal Press)

The Axe Book (pamphlet): Gränsfors Bruks, 2010 (Gränsfors Bruks)

American Axes, by Henry J. Kauffman, 1972 (Masthoff Press)

Early Logging Tools, by Kevin Johnson, 2007, (Schiffer Publishing)

To Fell a Tree, by Jeff Jepson, 2009, (Beaver Tree Publishing)

The Wood Burner's Encyclopedia, by Jay Shelton, 1942, (Vermont Crossroads Press)

A Reverence for Wood, by Eric Sloane, 1965, (Funk & Wagnalls)

Understanding Wood, by R. Bruce Hoadley, 2000, (The Taunton Press)

Holy Old Mackinaw, by Stewart H. Holbrook, 1948, (The MacMillan Company)

Once Upon a Wilderness, by Calwin Rutstrum, 1973, (The MacMillan Company)

INTERVIEWS:

Art Gaffer, Julia Kalthoff, Larry McPhail, Liam Hoffman, Mark Ferguson, Matt Bemis, Nick Zdon, C.W. "Butch" Welch, Harry Prouty

LOCATIONS:

The Division of Engineering Programs at the State University of New York at New Paltz

Hoffman Blacksmithing, Newland, North Carolina

Gränsfors Bruks, Northern Hälsingland, Sweden

Wetterlings, Storvik, Sweden

Mizuno Seisakujo, Sanjo, Japan

Diamond Machining Technology, Marlborough, Massachusetts

Tennessee Hickory, Loudon, Tennessee

US Forest Service, White Mountain National Forest Saco Ranger District, New Hampshire

Brant & Cochran, Portland, Maine

Stumpy's Ridge, Catskill Mountains, Andes, New York

Lorry Industries, Denton, Maryland

Cutts Island Trail, Rachel Carson National Wildlife Refuge, Maine

OTHER CREDITS:

The axe on the cover is an axe I found at Liberty Tool in Maine. The helve is hand-hewn. The maker and date are unknown. **The axe on the first page** is a Best Made Company American Felling axe, used by trail crews in the US Forest Service at the White Mountain National Forest, Saco Ranger District. The blade guard is made from tire inner tube and a discarded axe helve held together with duct tape.

Behind the scenes at Stumpy's Ridge, my field studio and product-testing facility in Andes, New York.

Acknowledgments

In some sense this book started a long time ago when I met Nick Zdon around a campfire at a design conference that I was speaking at in Northern Minnesota. I can't remember exactly what Nick and I spoke about that night, but surely it had something to do with design, the outdoors (and probably Jeff Tweedy, the lead singer of Wilco and Nick's doppelgänger). What we talked about would ultimately lead to Best Made Company and this book.

Over the course of the last decade, I sold a ton of axes, but the third axe I ever sold (a Famous Red) was to Nick Zdon. I sent him his axe in some of my earliest and crudest packaging, and in return he sent me a lengthy (and unsolicited) design proposal, detailing how I could up my packaging game. That marked the beginning of our work together. Soon enough, Nick came on full time to Best Made to be the architect of our customer experience program (including axe restoration), and so when I started this book he was one of my first calls. Nick has contributed directly to the making of this book. And you can see him in action on page 124. But Nick's contributions actually started when he bought axe number three. This book is essentially a distillation of our ten years of tinkering and curiosity-seeking, and it's my attempt to share what Nick, more than anyone, has taught me.

Ross McCammon has helped me with the words in this book, but so much more. Ross has been the creative coconspirator dreams are made of. Through his writing and editing, Ross elevated many of my most basic and convoluted thoughts (i.e., most of them). Through this project I may have taught Ross a thing or two about axes, but he has taught me much more about writing, and I am forever in debt.

When I started writing the proposal for this book, I had just left Best Made and was back where I had started: just me and a few dozen axes. I had to assemble a new team, and so I reached out to my agent, Nicole Tourtelot. Without Nicole there'd be no proposal, there'd be no book. Nicole was the first to believe in this project, and together we shaped it and she fought to get it in the right hands.

There was no righter hands than my publisher, Abrams. I am grateful to everyone at Abrams, especially my editor Rebecca Kaplan, and also to David Cashion and Michael Sand, who brought me on board. Their investment and belief in the project has been no better license to succeed.

Some parts of this book were easier to write than others. When I got to the Dynamics chapters, I hit a vast and seemingly insurmountable wall of science, engineering, metallurgy, and biology. And so I called Michael Getz.

Michael had started at Best Made on the retail floor. After reading his nightly store reports I promoted him to the back office, where he wrote copy and served as a product expert to our customers. Michael left Best Made and is now working on a PhD, but still found the time to contribute a vast amount of research and writing to this book.

Eleanor Hildebrandt came on board the book early. Before us she had been at *Popular Mechanics*, and we were very lucky to have access to her brain for as long as we did. Eleanor conducted many hours of research and interviews and brought order to a subject matter with a truly unruly history.

Peter Dudley's contributions to the book have been monumental, and took effect long before this book was ever begun. After about five years struggling to scale the production of our axes at Best Made, he took over and transformed our axes into refined masterpieces, and the workshop into a well-oiled machine. His talent and knowledge far exceed the art of adorning an axe.

Thanks to the great makers who have graciously let me into their worlds. Some of them I have known since I started Best Made, and others I met writing this book. I am especially grateful to Mizuno-San in Sanjo, Japan, Gabriel Branby and Julia Kalthoff in Sweden, Liam Hoffman in North Carolina, Grant Wanzer in Tennessee, and the guys at Brant & Cochran in Maine. Without them, there'd hardly be an axe with a story worth telling.

And I tip my hat to my team at Best Made; no one helped me build that company and tell the story of the axe more than them. I am grateful to my creative team: Maddy Tank, Stephanie Izzo, Katie Hatch, and Jason Frank Rothenberg. I am also grateful to John Maclean for representing the axe so well (and lending me some to shoot). And big hats off to Michael Laniak, a fellow Ahmek camper who braved being in front of the camera to show us all how it's done.

I also tip my hat to my earliest Best Made customers. One of the first was Francis Mallmann, no better champion, and adventure companion. And another was Chris Garby, an axe poet, and my gatekeeper to the trails of Maine and New Hampshire (see page 78).

I am in debt to Rob and Lisa Howard, and Hall Wilkie for their kindness and generosity, especially in the home stretch of this book.

I owe more to Meagan than anyone. She was there in the earliest days of Best Made and rode that beautiful bucking bronco with me all the way to the end. Now that's over and the book is done, Meagan, I am ready for the next rodeo with you.

A book has to have an author, and although I am proud to bear that title, I bear it reluctantly. This book would be nothing, and god knows where its author would be, without you all. Thank you.

—Peter Buchanan-Smith
Andes, New York, December 2019

WRITING, PHOTOGRAPHS, AND DESIGN: PETER BUCHANAN-SMITH

EDITING AND WRITING: ROSS MCCAMMON

RESEARCH AND WRITING: MICHAEL GETZ
RESEARCH: ELEANOR HILDEBRANDT

EDITORIAL CONSULTANTS: NICK ZDON, PETER DUDLEY, CHRIS GARBY

CONTRIBUTORS: MICHAEL LANIAK, JOHN MACLEAN, JARED NELSON, HEATHER LAI, GAZ BROWN, HARRY PROUTY,
CRISTIN BAILEY, MIKE KUHNS, LIAM HOFFMAN, JULIA KALTHOFF, C.W. "BUTCH" WELCH & MISS PEN

EDITOR: REBECCA KAPLAN
PRODUCTION MANAGER: ANET SIRNA-BRUDER

LIBRARY OF CONGRESS CONTROL NUMBER: 2020931085
ISBN: 978-1-4197-4767-0
EISBN: 978-1-64700-011-0

ABRAMS The Art of Books
195 Broadway, New York, NY 10007
abramsbooks.com

MIX
Paper from
responsible sources
FSC™ C144853

"Nothing matches the holiness and fascination of accurate and intricate detail."
- Stephen J. Gould

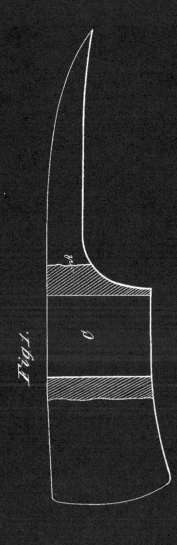

Fig. 1.

Inventor.
Harry L Lowman
Per Munn & Co
Attorneys

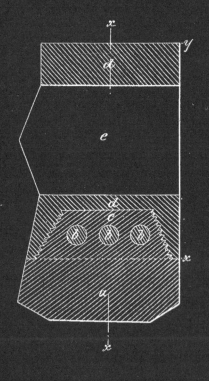

Inventor
John Lippincott